AF781305

THE DETA FACTOR

DOLPHINS * EXTRA-TERRESTRIALS * ANGELS

Adventures Among Spiritual Intelligences

by
Timothy Wyllie

Coleman Publishing
99 Milbar Blvd.
Farmingdale, New York 11735

Copyright © 1984 by Timothy Wyllie

Coleman Publishing
99 Milbar Blvd.
Farmingdale, N.Y. 11735

No part of this book may be reproduced by any mechanical, photographic, or electronic process, or in the form of a phonographic recording, nor may it be stored in a retrieval system, transmitted, or otherwise be copied for public or private use — other than for "fair use" — without written permission of the publisher.

First printing October, 1984
Manufactured in the United States of America
Cover design and illustrations by Timothy Wyllie

ISBN 0-942494-95-4

ACKNOWLEDGEMENTS

Any work such as this is bound to be a co-operative effort. First and foremost then my respect and gratitude to the 'consciousness unit,' Princess, Sandy, Christel, Paul, Krista, Hans, Nicholas, Ariel and of course, the Angels. Wholehearted thanks to our many friends who encouraged or participated in these adventures: To Armand and Rivka, Bobbie, Stuart, Christie, Stephen, Ruth, Jonathan and Terry, Cousin Chris and all my friends in England; to Estelle for re-introducing me to the dolphins; to Michael, Wendy, Charles and Daniel in sunnier climes; to Jaquie and Earl for solid support; to Wes and Jonnie for enthusiasm; to Oscar for sharing his story; to Sue and Derek for a chance to meet the dolphins; to Pola, Raymond and Freda for sanctuary; to Yanni and Larry for their encouragements in the last stages.

And to all Beings, creatures and animals, visible and invisible — this is really your story.

This book is dedicated
to the Creative Spirit, and to my
Father and Mother
and the wonderful
surprises in store for
us all.

CONTENTS

INTRODUCTION

We live in strange and uplifting times. The confluence of all of mankind's major conceptual streams fairly roars around us at this point in the history of the planet. Wherever we look we see the footprints of the Invisible World, but all too often, we find we do not have the appropriate belief systems, or even adequate perceptual apparatus, to appreciate and understand the extra-ordinary changes the Invisible brings with it.

Three hundred years down the line, they, or we as the case may be, might well look back and wonder how it all happened. How exactly did a belligerent and rebellious race slip so naturally into an era of Light and Life? Whoever saw it coming amid the seeming chaos of the latter part of the Twentieth Century? And more significantly for those of us living through the Great Transformation, just how were our consciousnesses expanded and extended to encompass the new energies sweeping the face of the planet?

This chronicle was originally conceived by my companion and myself as an exploration of dolphin intelligence, a perennial metaphor for the Wisdom of the Heart, but the deeper we submerged into the vastness of the mentation of these glorious creatures, the more we found the journey taking us into new and uncharted territory. We realized we must have been developing new physical and spiritual sensitivities **as** we were becoming involved. Frequently we were flying by the seat of our pants! We had to learn, through direct experience, how to remain calm and casual while talking to an extra-terrestrial presence. We had to understand new ways of listening to the quiet whispers of telepathic communication, and be able to separate them from the general hubbub. We had to come to terms on several occasions with whether we were going crazy or not. We had to absorb, through sometimes shattering circumstances, the reality of thought-forms, to know that thoughts have power. I have had to learn for myself how **not** to get stuck in ecstasy but to move through to the dimensions such profound bliss is designed to protect.

The DETA Factor turned out to be an experience which challenged some of our most dearly held truths about the nature of Reality itself. There will be some doubtless, who will reject the events reported as imaginary or fanciful, possibly I might have myself had I not undergone the experience. I believe what I witnessed happening to be true. I certainly would not have gone to all the trouble of recording it if I did not believe it so. I present it therefore as a chronicle of events which you, the reader, might simply allow to percolate within you: Allow it to resonate with your own Spirit of Truth and you will be able to detect its truth-content from the mass of mis-information to which we are all subject.

As becomes apparent through the development of The DETA Factor, the quest for dolphin intelligence opened up into other paths, all falling under the aegis of the wisdom and workings of the heart, and all showing the inevitable Oneness characteristic of life on higher levels of reality.

Through mankind's apparent nuclear capabilities, we are finally achieving a state of global awareness, a real sense we are all in this together. This will give way to a larger understanding still; that we are Citizens of the Universe and have an active, participatory role to play within the Affairs of the Universe.

Our migrations into space inevitably must be preceeded by explorations and expansions of our Inner Spaces. When we move into and beyond our solar system, out into the beautiful, wide, populated Universe, we will not be a race that bears such a burden of contentious aggression — all that will have been rooted out and left behind.

These have been troubled times, purifying processes invariably are. But in the deepest sense the worst is over. The corner has been turned; critical mass has been reached and the race will never be the same again. Two Hundred Thousand years of cosmic isolation are coming to an end. Conciliations have been made on higher levels of spiritual reality and this new peace is starting to filter through to us. A new feeling of profound hope is upon us which will give us all the courage and determination to break through the rational, materialistic world-views that have sat so sternly on the joy inherent in our hearts and souls.

The DETA Factor seeks to chart new territories of reality by describing the wanderings of two pilgrims of the imagination. We trust our experiences will be of value to others who enjoy the luxury, in this civilization of diversions, of following the subtle whisperings of their own hearts.

PREFACE

In the early days of receiving the revelations recorded here, both my companion and I found we had accumulated a sizeable conceptual investment in the rightness of what we were experiencing. This feeling appears to be a concomitant to revelation and simply has to be gone through. We are, after all, continually warned about the dangers of fanaticism.

Now, as I write some two years later, I have begun to balance the various emotions stirred up by the sudden reality shift I underwent. The voice of the man who started keeping notes on the odd things happening around him, and that of the man who now writes, belong to two different eras. The underlying message of reconciliation is working its wonders in my life.

That we all possess a good and a bad side to our natures is self-evident whether or not we have learnt negative aspects can be held in the realm of the imagination; and that these apparently opposing forces in each of us direly require reconciliation is, surely, the hidden lesson of our Age. If we are at war within ourselves, how can we ever stop externalizing this conflict, creating mutual hostilities and ultimately the devastation of war itself?

In truth these opposing energies have to finally meet and make their peace within the heart of each one of us. I have been drawn through my own training to perceive these energies as Gabriel and Lucifer, but they could just as easily be acknowledged as Yin and Yang, Ahura Mazda and Angra Mainyu of Zoroastrianism, the reference beam and working beam of holography, or any other expression of the dualistic interplay. Ultimately these are all metaphors; models we create for ourselves by which we can start to comprehend our magnificent Universe. And all credit to its ineffable richness that we can do this! In the reconciliation of these elements, these personalities, lies the growth of our souls. The ancient dualistic systems of thought all knew this. They were experiential processes in which the balancing of Light and Dark within the Initiates' life became the Opening of the Way.

Now, as an Age, we are moving towards assimilating the ancient knowledge as we ready ourselves for the tasks ahead. As we reconcile inside so will the Spirit of Reconciliation be carried around the planet; as we unite the polarities of a dualistic era, so also do we discover the triune nature of the New Age, the Age of God. As we learn to dance the triple-braided spiral path, so also will we enter the Era of Light and Life. For we are truly the hosts and stewards of the Planetary Soul, the Horus Child of the New Millennium, whatever the terms we use to describe this miracle. The feeling is the same. And it is **feelings** all sentient races share . . .

That I am I
That my soul is a dark forest.
That my known self will never be more
  than a little clearing in the forest.
That gods, strange gods, come forth
  from the forest, into the clearing
  of my known self, and then go back.
That I must have the courage
  to let them come and go.
That I never will let mankind put
  anything over me, but that I will
  try always to recognize and submit
  to the gods in other men and women.

Credo: D.H. Lawrence

CHAPTER ONE

The sea ran high, lightning crackling around me as I swam in my amateurish breast-stroke straight out into the teeth of the storm. My arms were starting to weaken, breathing coming in short, sharp gasps, swallowing mouthfulls of salt water, spray from the crests of the waves slashing at my face. It was now or never . . . I knew the dolphins were down there somewhere, I'd seen them before I set out on this absurd swim. Where were they now I needed them? Would they rescue me as I'd hoped and trusted, now that my legs were starting to cramp up on me . . .

But I'm getting ahead of myself . . .

I'd just turned 41 when I first met the dolphins. In a sense it was scarcely surprising since my life, on looking back on it with all its inevitable oddnesses, led surely and inexorably to these strange encounters.

Indeed, there are moments when I feel all my previous experiences were merely an elaborate training for the events laid down in this book; events which I am describing with as much precision as can be brought to bear on areas in which subjective and objective become of necessity blurred.

My training back in the 1960's had been in

architecture. Acknowledged as an excellent mixture of creative and practical, this discipline perhaps more than any other brought me in touch with the increasing inundation of our intuitive faculties by the groundswell of materialistic concerns. Still, I completed the long seven-year training with a degree more, I suppose, to fulfill the obligations of family and a government whose generosity had allowed me, through grants and scholarships, to pursue what even then was showing itself as an iconoclastic streak. The buildings I designed in the latter part of this training strained the practical vision of all but the most daring of my professors. The piece de resistance in my penultimate year at London's Regent Street Polytechnic caused something of a split in both faculty and students. It was a theater designed, through a process of computerized feedback and wrap-around sound and visuals, to produce a state of lucid waking dreams in the audience. I had researched cybernetics, still a science in its infancy, computer design, neurology, the workings of the eye and the psychology of hallucination. Looking back over the twenty intervening years I am still astonished at the relative sophistication of the concept, fuelled, as I now see it was, by the first of a number of shattering personal revelations. This first exposure to the nature of the 'invisible world' set me to questioning the very nature and substance of what I, at that time, rather naively considered to be 'reality.'

It occurred in the Spring of 1963 and I had been invited, I recall, to a party given for the American ex-patriot community in London by the jazz singer Annie Ross. A friend and I, rather unwisely I was later to think, swallowed five hundred morning glory seeds each before setting out for the gathering. It was not the first time I'd taken the hallucinogenic seeds — this was, after all, the 1960's — but I am certain it was the last!

On arriving at the party and feeling thoroughly nauseated I retired to the bedroom where I collapsed on the blue pile carpet. Soon the sickness passed and I opened my eyes. There, coiled within four feet of my face, was a large black and gold snake with glittering eyes and a flicking tongue.

I was immobilized with fear as the snake lazily unknotted itself and slid slowly towards me. We were both transfixed, the snake and I. It was as real and solid as the book on which I am now writing. I could see the pile of the carpet separating under its long, sinuous body: the shine of its vertical pupils. I felt the lightness of its breath and its tongue playing over my face. Then, in one continuous movement, it pushed itself into and through my right eye. I could feel it in my head, its thick body still coiled on the floor thrusting itself further and further into me. Then it was pushing and wriggling its way down into my throat and stomach, curling there for what felt like an eternity before reversing itself and forcing its head and body out through my now wide-open mouth. I have mercifully blocked out whatever followed, but I have some trace memory of becoming a nest to this family of snakes; of them entering and departing from my body at will.

When I came to the next morning my hair had started turning white. I was 23 years old.

The question I was left with, and in many ways still ponder, is if that snake was indeed a hallucination, then whatever is the true nature of what we've agreed to call 'reality'? And, in another sense, I'd had my first meeting with the Guardian of the Threshold.

The second experience I now regard as a turning point happened in the clinging climate of Nassau in the

Bahamas. It was the first of my religious revelations and, considering I had to that point been an out and out atheist, it came as a complete surprise to me.

In 1965, a year or so after completing my architectural studies in London and while working rather unhappily in an office in which the chronic boredom of detailed working drawings ground down on my spirit, I chanced to remeet a close friend of mine from the early college years. Robert had become disenchanted with architecture after three years of study and had left college to follow other drummers. He was some years older than I with a wider experience of life due to completing his National Service with the British army prior to entering architectural school. Well read in philosophy and psychology, we'd become immediate and lasting friends, so his departure and subsequent disappearance from my life had been a sad loss. We'd been a fine team in our time, wresting control for instance, of the college architectural magazine *Polygram,* from the editors one week before printing and, because we disagreed with current editorial policy, completely rewriting it in a fervor of sleepless activity.

And now here he was, back in my life, with a new wife and a self-help psychotherapy they'd dreamed up between them. It was all very new and it **was** the questioning '60's! Would I be their guinea pig, they wondered?

I was getting bored and disillusioned with architecture, turned off by the monumental egotism and endless attention spans required to see a project to completion. It was the right time and place and I fell innocently and happily in with their plans.

'The Process,' as I had originally named it, grew in size and complexity, helped largely by the infusion of my circle of friends. We were a bright and arrogant lot, sure our methodology would revolutionize if not the world, then certainly England, the small island on which we lived. How innocent we were! But, Lordy, we tried! Up on soap boxes at London's Speakers' Corner: talks at Oxford University and the London School of Economics. All met with ribald rejection. But to us, by now a firmly committed group of 30 to 40 young men and women, it simply proved how right we were and how sadly deluded was the world.

Then came our Big Moment. If the world wanted none of our idealistic pratings, then we would leave the world to its nonsense and head off to a tropical island with our community and start all over again. Island. Ah! Yes! Now where were there islands to be had? The Caribbean, of course. So it was lock, stock and barrel up and off to start a new life. Goodbye forever to family and concerned friends. I even sold my most treasured possession, a pre-war Martin guitar of inestimable value, to raise money for my ticket. Ridding myself of all worldly goods cleansed my soul, I told myself. Perhaps it did. Perhaps even then I was setting myself up for the revelation which was to change my life!

We were not at this point overtly religious and even if some members of our little group held God in their hearts I was among the still recalcitrant, believing, in spite of the snakes, only what I could directly experience with my five senses. Yes, I was certainly setting myself up for a very firm kick in the pants!

Then came **the** evening. We'd been sitting around quietly on the small patio of the fine old colonial house we'd rented in Nassau. We'd all taken jobs in the local community to refurbish our wilting funds for the great island-purchase which we convinced ourselves lay ahead.

I was sitting with my eyes closed, not thinking of very much, quiet but very much awake, when suddenly, to my utter astonishment, I found myself being hurled along in a mighty and surging flood of water. I opened my eyes in horror. The river continued unabated. I was struggling for breath. For survival. Thrown this way and that. Bouncing off massive boulders. Until I could fight no more and was sucked under to the fluid, cool depths below the raging surface. I believe I drowned at that point, I can recall no more.

The entire incident could have only lasted seconds yet it felt like an eternity. When I was washed ashore I was back with my friends, tears pouring down my face. I knew I had finally met the unstoppable force of God. There was no questioning the omnipotence or the omnipresence — I felt it in every cell of my body. I had battled and lost and finally had given myself over to the inevitability of a Presence so vastly more encompassing than anything I had ever imagined that I cried in ecstasy for the following three days.

Looking back on that undeniably powerful experience I realize it was my very rejection of anything I could not directly sense which forced the transcendent reality to communicate with me so directly. Here was something, after all, which I felt with **all** five senses. I was wide awake. I had abstained from hallucinogens for some years and, believe me, it was no acid flashback! It was overwhelming and as undeniable as poor Saul's journey to Damascus and it was my first introduction to the fact that something very remarkable exists just the other side of our normal perceptual capabilities.

After the experience with the river I remained with The Process for the next ten years. We travelled widely, finally finding our so-called island off the coast of Yucatan on the Gulf of Mexico. We lasted six months there, living off coconuts, fish, and in the end, famine relief, after almost being washed into oblivion by the great hurricane 'Inez' of 1966. We'd elected to stay and face the storm on the place to which we felt God had led us. Sure enough, Inez, having toppled the high walls of the ruins we'd made our home, turned a mere five miles from the shore and ravaged Vera Cruz instead. It was a remarkable lesson in faith and one which held the community together for much longer than we really deserved.

I spent the next three years travelling throughout Europe paying my dues on the highways and byways of a continent which cared little for the message of doom and destruction we felt impelled to disseminate. By this time it was the late '60's and disillusion hung like fog over the passing of the Old Ways. Like almost all evangelical communities we measured our success by the passion of the rejection we encountered. It was a harsh and bitter period which, like the wars of our parents' generation, seems considerably more fun in retrospect than the reality of those cold nights on Bavarian park benches or dew-encrusted Sicilian fields. We travelled with no money, merely the clothes and cloaks we wore, our faith that God would care for us and our large German Shepherd dogs who accompanied us everywhere. The adventures, of course, were endless and varied and would make a book in themselves. But it was still a period of preparation.

By the early '70's we were back in America and the era finally caught up with us. The enthusiasm and hope of the 1960's had given away to chronic pessimism and at last our miserable message of the doom-to-come was being heard. We busied ourselves with soup kitchens for the destitute, detoxification for the addicts and alcoholics, care for the aged and retarded, music and theater for those in prison, and so on and on. The bottomless barrel of broken and dejected spirits.

We were still paying our dues but it practically broke our spirits too. Perhaps it was the healing that saved us. Spontaneously this strange phenomenon emerged from our ranks, certain among us finding a simple laying-on of hands could bring about considerable relief of pain and suffering. There were even some quite miraculous recoveries. Possibly, we wondered, it was this all the preparation had been about, the training of a group of dedicated spiritual healers?

At this point, I guess, it all went to our heads. From having a number of highly effective but low-key Centers around the country, we all converged on New York City and bought an enormous building on First Avenue and 63rd Street. We were crazily over-extended, having to raise something like $40,000 a month just to keep going. Something had gone very, very wrong and we all knew it in our bones. The dark side of our group mind had gotten the best of us and the ambitions of a few managed to lock us into a situation from which there was little relief. The trouble was, spiritual healing and money do not, and should not, mix.

It was unfortunately a well-charted path downhill. The doom and destruction we had been so vociferously expounding started slowly and surely to rebound on us. We were working 20 hours a day, month after month, simply to keep this wretched status symbol of a building from the bailiff's hands.

Somewhere in among all this I had the next of my more startling experiences. I died. It was for me quite incontrovertible, although I didn't fully understand what it all meant until I came across the researches of Drs. Mooney and Elisabeth Kubler-Ross some years later. Here's what happened.

My body just broke down. I had walking pneumonia; my back had given out and I was terminally exhausted from attempting to hold together a disintegrating situation. If I managed three or four hours of sleep a night I reckoned myself ahead of the game. No holidays. No weekends. Constantly having to prop myself up; put on a good face; encourage a wilting and depressed crew of about fifty people to yet greater efforts.

One evening I collapsed. My body would go no further. The donkey could be beaten no more. I dragged myself back from our First Avenue monster to the relative peace of our house on East 49th Street and drew one final bath. I knew I was finished but I had little idea of what lay in store for me.

Within moments of stretching out in the bath, I found myself to my utter amazement, hovering somewhere out in space, my body clearly visible in the bath far down below me. The next thing I knew I seemed to be in a valley as real and solid as any landscape I have seen in my travels.

A monorail car was sweeping down towards me on a single, shining curve of metal. Then, mysteriously, I was inside the monorail cabin together with nine or ten other people. I can see them today in my mind's eye; opposite me sat a black man playing a trumpet with great beauty.

Somehow at that moment I **knew** we were all dying at the same time. A voice came to me over what I took to be a speaker-system, although it may well have been directly into my mind. It was very clear and lucid and quite the most loving voice I have ever heard.

"You are dying," the Voice confirmed for me, "but we wish you to make a choice. You can indeed pass on to what awaits you on the other side . . ." At this point I was given to see my body very casually sinking under the water of my bath somewhere below me. A simple and painless death.

". . . or you can choose to return to your life. We wish you to know, however, that you have completed what you came to do."

The Voice was utterly without attitude, wholly kind and considerate and with no bias whatsoever as to which option I might choose.

I thought for a moment with a crystal clarity I have never since experienced, and knew in my heart I desired to return to the world. On announcing my decision, there was an expression of delight so profound the monorail cabin dissolved around me, leaving me, once again, suspended in space, this time before a seemingly endless wall of angelic Beings. Such music and singing welled around me as I have never conceived, or perceived, before or since. I disintegrated into the overwhelming beauty of the sounds.

The next sensation was becoming aware I was standing on the edge of a vast, very flat plain. Beside, and slightly behind me stood two tall Beings dressed in white, or simply creatures of light — I couldn't see clearly, since my attention focussed on what lay in the center of the plain. It was an immense structure I can liken only to an extraordinarily elaborate and beautiful off-shore oil rig. It shone with gold and silver and had at each corner the faces of people and animals. Somehow it was in constant movement, yet in itself, it did not move. Intuitively I knew at that moment I was seeing the same structure Ezekiel describes with such elegance, although I find it difficult to retain the image now in my imagination.

I was led into this enormous place and taken to a brightly lit room where I was gently laid out on a flat surface similar to an operating table. Beings clustered around me, murmuring soft encouragement. Some apparatus appeared and I have a dim recollection of being hooked up to it. There was a moment of intense pain, except it was not truly pain. It wasn't quite shock either but some combination of both sensations. I felt as if in some way my blood had been completely changed, as if all the tired old vital fluids had been switched in one infinitely rapid moment.

I recall now only that after the "operation" I was taken 'somewhere.' I am still, to this day, rather embarrassed to admit I believe it was heaven. I have no conscious recollection of this last journey, my only clue being a strange sense of familiarity with a description of 'heaven' I came upon some years later in Robert Monroe's book *Journeys Out of the Body.*

The next thing I was aware of was descending gently again to my body still propped in the bath, the water now cool. When I got out to dry myself off I found myself completely healed, my back straightened and the thick phlegm in my lungs gone.

I continued working with the Community, now renamed The Foundation Faith, for another four years, and although we never did really achieve any kind of effective solvency we were successful enough to stave off the creditors.

The main instruments of this financial resuscitation were a whole series of conferences, courses and classes in spiritual development, and regular psychic fairs. I guess we might have even been the first to gather together a whole barrelful of psychics for the education and amusement of jaded New Yorkers.

The conferences were of lasting value and brought together experts specializing in parallel but differing areas of knowledge who themselves gained enormously from interactions that the more tunnel-visioned academia prohibits. The subjects were varied: Alternative

Medicine, Spiritual Technology, Tibetan Buddhism, Unconventional Methodologies of Cancer Treatment, but all allowed me the inevitable broadening of vista gained by hobnobbing with some of the top minds of the world.

But this period too had to end. By 1977 I had met, as they say, the Buddha on the road, and killed him. This time of training and preparation had drawn to a close and like many of us who have dedicated our lives to a guru, a spiritual community, or any external symbol of authority, I had to break free of the trappings and re-establish my links with the Divine Spark which resided in my own heart.

There followed a difficult three years. Leaving a religious cult is comparable, though probably even more painful, to leaving a long marriage — and I'd been involved with the Community for over thirteen years. Twenty four hours a day . . . seven days a week, with little or no respite.

It took a year of solid talking, of self-deprogramming, of coming to terms with the multiplicity of choices suddenly facing one who has lived vows of poverty, celibacy and obedience, to realize I was truly my own man. The insidious nature of 'organized' religious systems became a whole lot clearer to those of us who had made the break. The hypnotic holds, the manipulation of guilt and the presumption of superior knowledge wielded by the majority of the priest (and priestess) caste left me determined to follow my own inner guidance. It was a long, and in many ways, harsh lesson, but one which indelibly prepared me for the new dimensions of self-government which lie ahead for all of us.

During this period of detachment I started a small business in New York, marketing a system for keeping color slides and photographs in good and readily available condition. I'd had a hand in the development of the system some years earlier and the timing now rendered it appropriate to introduce it to corporate America. After everything I had been through it was a splendidly down-to-earth activity which provided me with the chance to jump into 'normal' life and re-assimilate what was of value in the everyday affairs of men.

Though it was a difficult time for the economy, the business went relatively well — well enough for me to turn the work over to a distributor within three years.

It was sometime towards the end of this period that the new trail started. And it started, as these journeys into the Inner World often do, with an odd chain of meaningful coincidences, of synchronies, as Karl Jung named them. All I did was join John Lilly's Human/Dolphin Foundation and, within two weeks, it was dolphins every way I looked! An old friend, not seen or heard from in years, turned up out of the blue with stories of her dolphin center in New Zealand. She was pioneering underwater birthing using the dolphins as midwives.

Next, it was two short movies about dolphins, one of which was directed by, coincidence again, another old friend, Michael Wiese. The other, of a rather more academic and investigational tone, contained an incident which, though it was tantalizingly fleeting, raised in my mind more questions than Michael's superb underwater camera-work. I will relate it as carefully as I can since it deals with the immensely significant possibility dolphins are telepathic. But before that however, a word on what telepathy is or might be.

Simple thought transference, or the psychic picking-up of images in laboratory testing situations has not, as any scientist will tell you, yielded results more than marginally convincing. Dr. Rhine and his hapless volunteers spent most of the 1940's fruitlessly peering at cards with different symbols on them and ended up proving absolutely nothing. Other more recent testing procedures have been only slightly more encouraging. Yet the subjective experience of telepathy persists.

Who hasn't got a friend or a relative with a story of someone saved from a disaster by a last minute 'hunch'? How often in our own lives have we an intuitive conviction something is just so when the only way we could have known it involves telepathy?

No, there is no doubt our race **is** telepathic, although it seems to manifest most vividly in moments of extreme need. But have we therefore come from a telepathic past, the quiet whispers now buried under a barrage of mechanical sounds and electronic signals? Or are we heading into a future in which we'll accept telepathy as one of our normal senses?

And what of the dolphins? Might they have developed some form of telepathic awareness along with their admirable sonics? Being able to detect elements of differing densities by bouncing sounds off them must allow an ever-shifting pageant of the most intimate information. Glandular changes in the bodies of all mammals reflect variations in emotional and physical well-being, and dolphins, with their long history and sensitive accoustical systems, are undoubtably able to gauge their companions' welfare with an accuracy unknown to us. Whether this comprises telepathy or simply an acutely empathic awareness I did not find out until I knew the dolphins a little better. And, if we scrap the idea of telepathy being strictly **thought** transference, and widen our parameters to understand it is the nature and content of the thought itself, its emotional reality, which governs whether or not the thought is worthy of being 'transferred,' then it makes it much easier to see why the scientists working in cold laboratory settings will never produce convincing evidence.

Now we come to the second of the two films I'd seen, this one by British marine biologist Horace Dobbs, and an encounter which could well be revealing a telepathic ability in action.

Dr. Dobbs had made a small wooden surfboard which is being towed behind their motorboat. His female co-worker and the cameraman are in the boat and have been pulling first Dobbs, then Donald the dolphin, who has nudged the biologist off the board to play with it himself. Donald's presence in the drama is, in itself, somewhat strange. He's appeared in the cold waters off Britain, already an unusual occurrence, and had quite freely made himself available for whatever study the scientist devised. There was no question of Donald being held in captivity yet he remained with Dr. Dobbs and his team for over three years. All very odd! Almost as if the dolphins had set up the appropriate link themselves . . .

Back to the film: We're sitting in the boat which has left Dobbs some two hundred yards further back in the water, when, after a few minutes of tugging Donald along on the surfboard, the female researcher evidently thinks it is time to stop the game and pick the scientist up. She starts speaking.

"We must go back and pick up Hor . . .," and then she continued, ". . . oh, I see Donald is already on his way!"

And indeed, by the time she got to the name Horace, Donald the Dolphin, who was at least twenty five feet behind the boat, had turned back towards the black, bobbing head of the scientist.

There is no possibility Donald could have heard her talking in any conventional sense, the distance was too great and her voice an aside to the cameraman. Granted it could have been the dolphin's anticipated response;

maybe he always turned back to the person on the board. But there was something in the woman's tone of voice as she completed the sentence, a curious mixture of tenderness and embarrassment, as if she'd moved in and out of a trance state and felt a bit raw at being caught so open-hearted.

It all puzzled me at the time yet it was this odd little incident, not even commented upon in the film, which led me to feel intuitively there was a strange and wonderful mystery to all this.

A further coincidence of events came up as I was mulling through the thoughts generated by the sudden influx of dolphins into my life.

Some friends of my companion called from Florida. I had never met them nor had they any knowledge of my recent interest in cetaceans but they happened to mention in passing that dolphins had taken to leaping over their boat whenever they'd gone out. It must have been relatively unusual for them as my companion relayed to me their excitement. It was a synchrony too obvious to miss, and they happily extended an invitation to both of us to come and stay with them. That gave me a week to bone up on what is generally known about cetacea, the family of mammals of which dolphins are a member.

I found to my surprise relatively little is actually known about these animals. Sure, there's been a lot written, but it either repeats itself or is a mass of anatomical data based on postmortem examinations. Although an acknowledged leader in the field, Dr. John Lilly's work has been at something of a standstill since he started trying to find a mechanical way of talking to the dolphins. He is also working with captive dolphins and I had already sensed the possibility their delicate brain functioning is impaired by the trauma of imprisonment. Some species of open-water dolphins, or so I'd read, have such minimal a concept of physical boundaries they will batter themselves to death against the walls of a tank.

Unlike us and our land-lubbing love of territory, the dolphins have a history of over thirty million years of physical freedom. Freedom in three dimensions too! It is more than possible even those surviving in captivity have little concept of physical boundaries until they are locked into a small pool. I wondered at this point whether this explained why much of the more innovative research work in the field is being conducted with wild dolphins, like Donald, who have approached humans spontaneously and uncoerced.

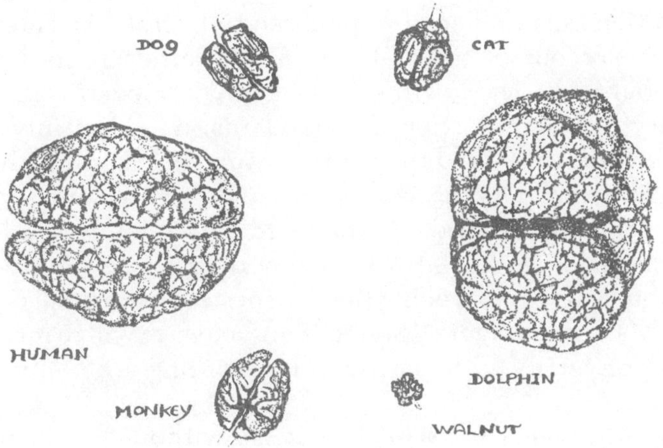

BRAINS

Any speculation on the nature of dolphin intelligence ultimately has to focus on those wonderful, large brains. The evolutionary design process certainly does not bestow organs as complex as those on mere whim. They are there, so the dolphins must use them. But, whatever for?

Some experts suggest movement in three dimensions requires more complex neural networks. While I am sure this is broadly true, birds manage pretty well with their proverbial bird-brains. Others will cite the possibility of magnificent memories; after all, whale songs are long and complicated, something like reciting *The Iliad,* says Carl Sagan, in 90 minutes, then repeating it phoneme-perfect all over again. But this somewhat sidesteps the issue. What could they be singing about? The Sagas? Their aural histories? Surely more than a weather report!

Then again, recalling my experiences in an External Sensory Withdrawal Tank following in John Lilly's noteworthy tracks, I became more aware of the kind of conditions so familiar to dolphins and which allowed me a glimpse of 'out-of-the-body' realms. Whether or not the experience of leaving the body is given credence as a reality by current scientific standards, it is certainly a **subjective reality** and therefore presumably requires the neural complexity necessary to handle these journeys into the Inner Worlds. I wondered if the dolphins and whales had developed a mastery, over those long 30 million years, of travelling the dimensions of inner space? The Master Voyagers, leaving their bodies at will to meet and converse with entities frequenting the Inner Worlds . . .

As you can see, I wasn't without rampant preconceptions when we took off for Tampa, Florida on a warm autumn day in 1981. Having looked through the literature and probed my own deeper feelings, I found myself going through a preparation for a meeting with sentient Beings of intelligence, but of a totally different form of consciousness. If there was to be a mutually shared arena of contact between us, **and** if it turned out my new dolphin friends would be even halfway as consciously intelligent as I speculated, then I'd better have some good conversation at my fingtertips!

Consequently I came up with five questions either of general planetary relevance or, as in the case of question five, of a more personal interest. In the esoteric tradition, I formed and held these questions in my mind and then, after a day or two of consideration, simply stopped thinking about them.

The questions I carried with me to Florida were as follows:

1) How do dolphins deal with violence and predation?

2) How does a complex and sophisticated society (like the dolphins?) handle disease without the apparent use of technology?

3) How does a complex and sophisticated society sustain a sense of continuity without apparent instrumentation, artifacts or extra-somatic memory systems (books and the like)?

4) How do dolphins balance their populations?

5) Since UFO's consistently have been reported entering and leaving our oceans, have the dolphins had any contact with them?

Timothy Wyllie
2/6/81

CHAPTER TWO

On arriving in Florida my companion and I spent that first weekend at our friends' house talking and unwinding from New York's psychic pressures. One short trip out in the boat had yielded no sign of dolphins and now the weather had set in.

Came Sunday and a neighbour's child reported seeing three dolphins swimming in the small bay behind the house, apparently considered by general concensus a rare event. We calculated the dolphins must have come within thirty feet of where we'd sat talking. It was our first and tenuous contact.

Our friends, bless them, then put us up in a sumptuous fourth floor condominium some miles from their house. It overlooked a long, gently curved white beach and after the weekend discussions were over we started our explorations.

Deciding to follow hunches wherever we could and not knowing when or how the dolphins would make themselves known, we relaxed and enjoyed ourselves, going for a swim in the late morning.

We strolled down the beach toward a monolithic pink cluster of Disneyworld minarettes — the San Cezar Hotel, sitting like a mosque for sunworshippers on the reclaimed fills of the south-western shoreline.

A blonde middle-aged lady whom we passed pointed out some pelicans bobbing in the bay, and told us dolphins might well be following the same shoal of fish.

"And when you see dolphins you seldom see sharks," she added, anxious for our serenity. A good piece of folk wisdom, and gratifying to see it becoming more generally known.

One hundred and fifty yards further down and well before the Mosque we settled onto the sand. My companion was later to say she started **feeling** the presence of the dolphins at this point. She also felt very 'drawn' to sit at this particular spot.

Dozens of different types of birds, resting no doubt between meals, paced around on the sand between us and the sea. The sun shone in an azurite sky. There was little or no wind and the sea was oily smooth.

We walked, then swam, slowly and luxuriously out into the waters of the Gulf. Within a minute and a half we saw two dolphins feeding off the far side of the shoal some seventy feet away. These were the first wild dolphins I had seen from the water and I found myself in a near-rapture state of tears of joy within moments. We both started swimming towards them until we noticed they continued to keep their distance.

My companion stopped and swam pointedly in the other direction hoping the akido gesture of quiet play would draw them closer to us. No moves from them.

I then 'felt' as strongly as I could, making a conscious effort to project a challenge towards the dolphins: that they could probably cover the intervening seventy feet in . . . how many seconds? . . . starting from . . . NOW! One . . . two . . . three . . . four . . . five . . . six . . . and there was a resounding splash a few feet behind my back. My companion, not wearing her glasses, thought she saw a flash of navy blue out of the corner of her eye.

The movement was so fast, so unexpected, I felt the hairs on the back of my neck rise, with the fleeting impression that one of the dolphins had responded to my telepathic challenge before the thought was chased off by a somewhat safer rationalization — it could have been a large wave. (Sea was oily smooth?)

Then silence . . . moments later they were out there again, about seventy feet away. I decided there was nothing for it but to swim out to where they were feeding. Now, I'm not a strong swimmer and the prospect soon felt formidable. The dolphins too were keeping their distance, and before long there was no sign of the seabed beneath my feet and my arms were starting to ache. I calmed myself with the apocryphal tale of the old rastaman living on a deserted Jamaican beach. Each morning he swims straight out to sea as far as his strength will take him. Only then will he turn back!

Looking back on the experience of that swim I realize it was at this point I must have fallen into some sort of telepathic rapport with the dolphins. A mental flash of their saving drowning sailors seemed to come from another mindal source and layered over my imagination. It wasn't a thought that I generated. This was followed by a sense the dolphins wanted to rescue me; wanted to demonstrate both a kindness and what they feel they do best.

Did I have to swim until I panicked, I wondered? Something about that dramatic sacrificial act did not altogether appeal to me at this point and I tried signalling telepathically: we'd establish contact in another way.

By this time my arms were getting progressively weaker and I was some two hundred yards from shore, a long way for a slightly uncertain swimmer. Well, I knew I wasn't going to drown, because my likely panic would alert the dolphins into saving me. In fact I found I simply couldn't panic on spec., so I took to wondering why tired swimmers flail about when all they have to do is turn on their backs and float . . . and rest . . .

Now, where did that thought come from?

I turned on my back. My legs, never blessed by the buoyancy of excessive blubber, began to sink.

Then, to my astonishment, they touched the bottom and I was able to stand. Although still far from shore I'd come upon a sandbar. On it I rested for several minutes still chuckling at my imaginings and not realizing yet I was being carefully primed by the dolphins.

I continued after them, this time wading in the warm shallows of the sandbar, which I found on looking back at the beach, was the second of two sandbanks, both narrow strips running parallel to the shoreline. I must have swum right over the first one!

The dolphins had veered in towards the shore and I was almost on top of a small rubber dinghy before I noticed the two men and a boy diving for shells. My concentration was so focussed on the dolphins I was genuinely startled to see people.

The boy appeared from underwater holding a flat, brownish disc about four inches in diameter. He told me it was a sand dollar and I recalled having seen those bleached white mandalas — a cousin to the sea urchin — in friends' apartments without quite knowing what they were.

One of the men took to explaining the strong religious connotations of the patterns on the disc's surface. On one side he showed me the five-pointed star of Lucifer and Satan, on the other a poinsettia and the gashes, he suggested, were Christ's stigmata.

What happened next occurred with such clarity that I, who have the advantage of knowing what lies ahead, can only ask you temporarily to suspend

Common Sand Dollar ~ Echinarachnius Parma
(Top and bottom views combined)

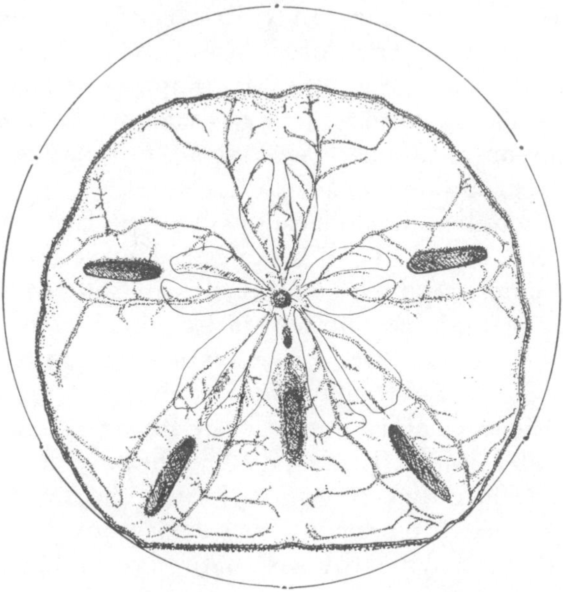

judgment. Moments of telepathic union are fleeting and certainly non-verbal. Yet if you have had one, you will **know** they exist and with what lucidity they illumine an issue.

This was one such moment. I'd been standing with the sand dollar cupped in my hands, sunlight wrinkling the ocean and I **knew** in one golden flash the dolphins had led me to precisely this place. I knew at that point I was intuitively in contact with a vastly intelligent species who had indeed used their large and active brains to produce an advanced, and what's more, **technologically** proficient society.

The visions were rolling in now, layers of images replacing one another as they came in clusters. I saw the dolphins, possessing no hands or opposable digits, creating artifacts by modulating sound waves. By overlaying living protoplasm with accoustic holograms they are able to store and retrieve information from living organisms. The closest I came to understanding it was in the sense of an organic computer — a shell, for instance, like a Nautilus, acting as a dolphin book, storing information until it is released in an information cluster at the appropriate signal.

Then the sand dollar. I saw how similar the surface patterns were to those symmetrical wave formations achieved in that simple experiment with sand and a metal or glass plate. If a violin bow is drawn gently across the edge, the sand vibrates into a symmetrical pattern very similar to the one I held in a frozen state in my hands.

The revelatory hit was so strong it seemed to bypass any intellectual faculty I might have erected as a barrier to intuitive conviction. I **knew** the dolphins had developed extrasomatic memory devices just like us but, by working in sympathy with natural conditions, they'd grown and shaped them.

I also knew I would have no way of actually proving this except by following through on faith and believing what I had seen was true. What I was being shown was well beyond the capability of Western scientific thought, yet if it were so, I would have it confirmed, in the way of the Inner Life, by continuing circumstances.

What I did not realize at the time was that the dolphins were starting to answer my five questions.

My companion caught up with me while I was still standing in the intuitive reverie holding the sand dollar. She'd been having her own adventures. She told me she had felt a bit left out seeing me a hundred yards away and sensing something dramatic was happening. The moment she allowed this feeling in she found herself surrounded by a shoal of tiny "finger fish," flitting and flipping around her. Wherever she turned they followed, jumping up into her hair and kissing her face. We wondered whether the dolphins had a hand in this, laying on a show for each of us in their own way. After what I'd seen, anything was possible!

Deciding to go with our intuitions, we held up the sand dollar, our new-found biological communication device, and pointed it at the quadrant of the sky in which the sun would be at about 5 p.m. that afternoon. We were both getting hungry — perhaps it was all the dolphin feeding going on — and wanted to fix a time for the next encounter with our friends.

The sand dollar stayed with us throughout the afternoon. At the very least it must have acted as a psychometric stimulus since we both became more attuned to the dolphins' reality. It was a great game; first one of us holding it and "showing" it to the artifacts of land-based life, then the other, talking and chattering away into it.

Through this process we sensed a number of clear impressions of the way dolphins perceive reality. They were no strangers to organic substances like trees and plants, but when we held the sand dollar up to metals and glass we both received a strong feeling of non-comprehension. Their sonar, evidently, is not powerful enough to shape inorganic material. Thus their problem in comprehending the nature of many of our artifacts. They seem to have the same vulnerability to cultural hypnosis as we do: what they have difficulty in perceiving they have not been able to conceive.

By 4 p.m., the promise of a boat we'd hoped would take us out to our pre-arranged meeting with the dolphins had fallen through. Using the sand dollar in the same manner as previously, we relocated the gathering in the inlet just behind our friends' house, where we'd spent our first evening.

With the sun approaching the previously agreed quadrant, we sat on a jetty extending out into the bay and kept our eyes open for leaping dark forms.

In a matter of minutes, an enormous thundercloud had built up behind us. It darkened the sky, causing great swirling masses of deep ochre storm clouds to gather and sudden gusts of wind to whip up loose debris. As we watched, amazed at its vigor and suddenness, the storm split into first two, then four, smaller storms which seemed to dance all around us. Yet on us, in the center of all this activity, it barely rained a drop.

Just a few minutes before this spectacle we'd been sitting on the jetty in broad sunlight being shown a superb Audubon book of underwater photographs by our host's three-year-old. Odd that too! He'd quite spontaneously gone to the house and stumbled out with this large book almost as big as himself. He was also surprisingly insistent we looked at it. Then the storm started and I put my jacket over the book to protect it.

When the storm had passed as quickly as it had built up, I pulled out the book again. Lo and behold, on the cover, which we had not yet noticed, a full color picture of a gently smirking dolphin.

Had this been their own sweet and tricky way of keeping their meeting with us? And what of the thunderstorm? Was weather control too part of their repertoire?

In the evening sun, we now turned the pages of the book with an altogether new interest. Exquisitely formed and illuminated jellyfish looking for all the world like the most opulent and baroque lamps; an enormous variety of coral appearing in detail as if it had been fashioned by the finest artists the planet had ever seen; reefs we read that "protected whole islands from annihilation by the elements" — shades perhaps of what the dolphins might be doing down there.

As we read, we both had the growing conviction we were looking at dolphin artifacts. The seas, we surmised, must be redolent with cetacean works of art, yet totally unacknowledged by our race. We were filled by the most profound pleasure in the possibility the dolphins could have created, in those long thirty million years, a truly graceful, compassionate, complex and playful culture. Not only have they a high measure of technical knowhow in the biological sciences, but they combine this with a boundlessly intricate and creative spirit.

There comes a time in any inner journey when a leap of faith is required. The agnostic is seldom prepared to make this jump, preferring the relative security of his sense perceptions. But this was different. We were being taught, shown through a whole variety of experiences the way another race lived. It could of course all be in our imaginations. Yet, as I found out considerably later, it is precisely **through** our imaginations we **are** available for contact from more subtle realms. And the way we feel about it is the surest guide to the reliability of the contact.

The objective scientist in me was starting to take a back seat to the more poetically inclined intuitions. I knew in those moments of inspiration that the dolphins, God bless them, are here to collaborate and enter into open exchange with their land-based brothers. That through contacts like the one they were establishing with us, they are demonstrating their many social, artistic, ethical, romantic, biological and communicative skills which they wholeheartedly wish to share. Their aim? To produce, through co-operative effort, a conscious race representative of all life forms on this planet.

No dolphins were visible throughout the next morning. We deliberately didn't think too much about the various options and possibilities open to us after our first encounter. We didn't want to develop an awkward level of self-consciousness, knowing it would block any form of telepathic communication with them.

Around mid-day we walked to what we took to be our spot on the beach, spread our blanket and settled down. Within minutes, the place was running with ants. By this time we'd established this was not our "place," but one extremely similar some fifty yards nearer to our condominium. Rejecting the immediate impulse to move, we talked to the ants as persuasively as possible to keep off our towels.

They were after the previous day's sand dollar which we'd brought with us. We'd put it down in the shade and within seconds it was covered with ants, possibly drawn by the smell and possibly, because no individual ant spent that long on the surface of the sand dollar, drawn by other reasons.

Ants, after all, would be a complete novelty to a water-based race like the dolphins, and, if there were biological specialists among them fortunate enough to hear whatever sonic vibrations resonated off the now ant-covered sand dollar, they must have had a field day of new sensations.

We bathed then in the warm gulf water, with a fair wind picking up the top of a choppy surface. No dolphins. No fish or birds either. We played and frolicked, feeling our joy would somehow communicate itself over unheard frequencies to our friends out there. We cuddled and kissed. The water moved around us, as we gently and happily made love, the marzipan minarettes of the San Cezar melting sedately in the background. Still no dolphins. A strong sense they are playing with us. It feels like they play everything. It's their thing.

We finally stumbled ashore and realized the spot we'd originally headed for was actually considerably better, and moved all our things the further fifty yards down the beach. I left the original sand dollar to the ants and we suspended a new one I'd picked up on a beach-grass with the tab from a metal beer can. As I was doing it I recalled my earlier thought regarding dolphins and metal. Some they surely know about — gold, silver, iron, and those occurring naturally — but the way we alloy certain metals with fire can only be a total mystery to them. We'd settled much more comfortably into our original power spot and within moments my companion pointed out two leaping forms about 35 feet offshore. A large and small dolphin breaching at exactly the same moment. Soon we saw eight or ten others. All leaping much higher than yesterday, right out of the water and quite evidently watching us.

We resolved we wouldn't swim out if they passed by. They didn't . . . they stayed opposite us. I elected to go after them. Always the sucker!

Of course, by the time I'd swum my tedious way out they'd disappeared again. So this time I made like I **was** drowning (and I nearly did in the process). Not a sign. I felt a bit deceptive too but I **had** put my heart into trying. It just seemed such a silly way to go, as I thrashed my legs about and shouted.

When I'd calmed down, I had an impulse to bring up another sand dollar. It had a large chip missing out of it, the only one I've seen partially damaged but living. I then received a sense to break it in half and I got a feeling of great joy together with the impression that I'd broadcast specific vital information — maybe the death throes of an organic walkie-talkie! And then again, the curious feeling something was using me for an experiment simply because I **had** opposable digits and could facilitate some sort of instrumental calibration. The impression faded.

I waded back to the shore with a little more thrashing for good measure. I recall shouting some sort of rather frustrated challenge down the sand dollar and felt a little stupid standing there talking earnestly into a sea urchin. It all felt so much on their terms! If only they'd close in a little and I could actually swim with them and touch them.

Minutes later another superb thunderstorm chased all but the two of us off the three-mile beach. Lightning crackling all round, and the rain, whipped into pellets by the wind, flung us happily back into the water. We had momentary qualms about being struck by lightning but, as the movie says, "we're going for broke."

We clung together in rising wind-slicked waves, the warmth of the water contrasting with the cutting cold raindrops. We made for the second sandbank, clinging to each other and feeling the pull of a new and ebbing tide. We finally reached the sandbank safely. No dolphins. "Gee guys, we coulda got killed!"

Nothing for it but to make love again. After all, what other more appropriate act might draw the dolphins to approach us?

The storm lifted slowly and then warmly. The sun poured out in a great shell of radiating colors. The light rain sparkled into a broad ephemera of a rainbow. Rainbows and dolphins. Dolphins and rainbows . . . They go together in some occult way . . .

The impressions started rolling in, washing over me and eliciting memories and new associations . . . I recalled how astonished I'd been when I found I had been able to disperse small clouds by concentrating energy on them. A common effect, apparently, that Itzhak Bentov mentions in his book *"Stalking the Wild Pendulum."*

Strange as it may seem, it actually works. Go out and try it. Just still yourself inside, select a small cloud and focus warm energy on it. Within half a minute or so it will have shrunk and disappeared. An evident example that consciousness **can** affect weather.

As I stood there, chest-deep in that deserted Florida bay, I realized I was receiving confirmation of yesterday's fleeting impression. Indeed the dolphins do have a hand in controlling and maintaining the weather. If my paltry consciousness could knock clouds out of the sky, how much more could the dolphins influence weather conditions, with their considerably wider faculty.

Writing in my journal that same evening I noted:

— I find myself again feeling as if I am a pawn in someone else's game. I have the overpowering sense these words are being drawn out of me — that they need to be communicated.

If I'd have had the close physical contact with the dolphins I'd hoped for, I wonder if I would be receiving the telepathic impressions in quite the same way. It is as though I was taking **their** intelligence test with the promise of swimming up close to them, the carrot luring me on —

Later there occurred a small incident which, had it happened under different circumstances, might have caused me considerable concern.

I'd gone to the bathroom towards the end of the evening; after I had finished peeing, and just before I pulled the handle, my eye was drawn to a fleck of blood sticking to the side of the toilet bowl. I knelt to examine it and found it was part of a four-inch fine thread made of what looked like a combination of blood and sperm. It was delicate and broke abruptly when I touched it. It was a little scarey and thoughts came into my mind of testicular cancer since I'd recently read Arnold Mandell's account of the disease.*

Oddly enough, the thought didn't have any holding power. It came and it went, leaving me with an unexpected feeling of joy and a conviction the dolphins had done something to me. I've never had a discharge like this before and yet I found myself curiously unworried. I was to find out the next day why.

We could see the dolphins in the bay that next morning from our condominium window. They were waiting for us . . .

We took our time before deciding to join them. Tricksters that they are, they'd disappeared from sight the moment we reached the beach. No matter, we knew they were out there and I wondered whatever they had in store for us today.

With that we took to the water, warm and calm again. As we lazed and swam in the strip of deeper water before the first sandbar we found ourselves talking in some depth about our relationship. Obviously every couple does this but here there was a difference. For reasons we'd yet to fully understand, we found our consciousnesses were being expanded as we talked. It was strangely similar to getting high, as if we were being rapidly upstepped by an external force operating at a higher level of vibrations.

I had a sense of our mythic personalities and saw how we are all, in a thoroughly natural way, a multiplicity of selves. How these selves, or aspects of the basic God-given Personality, are stacked like the layers of an onion. In those moments of clarity I saw how deeply each of us

* *The Coming of Middle Age,* Arnold Mandell

reached into the history of the Universe . . . the parts each of us have played . . .

I knew then the dolphins were replying to my five questions, but with some additional seasoning of their own. They were getting us to directly **experience** the answers in the way of good teachers everywhere.

I had my answer to No. 3, concerning dolphin artifacts, in the sand dollar. They'd achieved continuity by using their sonics to create biological libraries. What else might they have applied this extraordinary faculty to?

At that point, the image of my discharge the night before came up in my mind with its lurid but momentary message of testicular cancer. I broke out laughing when the truth of it struck me. Of course, the dolphins had zapped me with their sonics! I'd undergone a simple piece of ultra-sonic surgery and if, indeed, I had some internal imbalance then I was fully cured.

I had no way of proving all this but I was in a state of consciousness in which such things need no second opinions. It was also clear to me this use of ultrasonics is precisely how dolphins handle disease and, in all likelihood, how they practice birth control too.

That left my question about predation and hostility. I'd read how dolphins will drive off sharks, often killing them by butting them with their beaks. Apart from the occasional confrontation with a shark, I wondered if they had anything similar to the contentiousness we land-based animals experience.

Perhaps the next, rather unlikely, interchange served to illustrate to both races the dynamics of some human encounters.

As you can imagine, both of us were in a happy and inspired state as we made our way slowly back from the sandbar. Although we had not actually seen the dolphins we knew they were swimming close to us. At one point my companion felt the lightest of clamps around her lower calf. So distinct was it, she thought I might have been playing with her until she saw I was 6 feet away.

A small motorboat pulling a waterskier skidded past my head. It seemed somewhat close at the time but I was still too enraptured to be disconcerted.

It came back for a return run and, in case they hadn't seen me, I waved happily at the two young men in the boat. No response. I waved again. Still nothing, though I knew they'd seen me. One more time I tried but met with their dull, glowering stares. I turned the gesture into "the finger" and went back to talking to my companion.

Within a couple of moments, I saw out of the corner of my eye the boat making a sharp turn and heading right back towards us. The lad behind the wheel was screaming and shouting at me. He was obviously the bully type and evidently enraged by my gesture.

I tried apologizing. I'd only done it to express my frustration at their total lack of response but I hadn't bargained for this.

No way was he going to reason with me, let alone even listen to my apology. He was too far gone into his anger. He pulled the boat up to us, leaning down and taking great swings at my head, swearing and shouting until even his friend was looking at me in desperation. For some demented reason he wanted me to get into the boat so he could punch me.

The whole scene had such a surreal quality about it — he didn't seem to want to jump down into the water with me — **because** there was so little he could do but scream and rant.

I found myself, rather surprisingly, calm and confident throughout the encounter which must have lasted all of five minutes before he gunned his engine, still shouting murderously at my "Yankee" ways.

In the middle of this vivid interchange I became aware the entire episode had been staged for the dolphins. **By** the dolphins, even!

Underlying the grosser emotions, I detected a much deeper sense of regret. As though the dolphins were such peace-loving creatures even this relatively insignificant situation had brought them considerable pain. I realized then that dolphins have no real comprehension of hostility or contention. Their interdependence somehow precludes it. The irrational hatreds we humans think little of throwing at each other haven't normally fallen within their perceptual spectrum.

It was at this point I started catching a glimpse of the dolphins' point of view. A reason as to why they were going through such elaborate procedures to make contact with us. I had the impression the dolphins were starting to come to terms with the nature of man. They were having to make some very basic adaptations in order to understand us. I saw how they had broadened the base of their capacity to perceive negative 'emotional wavefronts' in order to make contact with us. This unwholesome encounter had been one such experiment. Throughout it the dolphins were having the chance to analyze glandular, chemical changes in us, the protagonists, and had thus moved closer to gaining a fuller appreciation of one of the quirks of our species, surely one of the major stumbling blocks to their understanding of us.

The answer to my fifth question, about U.F.O.'s, came on our final evening in Florida. As with all our encounters with the dolphins the circumstances surrounding the event were synchronous and significant.

I had been sitting on the balcony overlooking the

~UFO Sighting~

· 21h02 - 21h22 · August 25, 1981 · Near Tampa, Florida. ~

First Manoeuvre: 8 minutes.

Second Manoeuvre: 12 minutes.

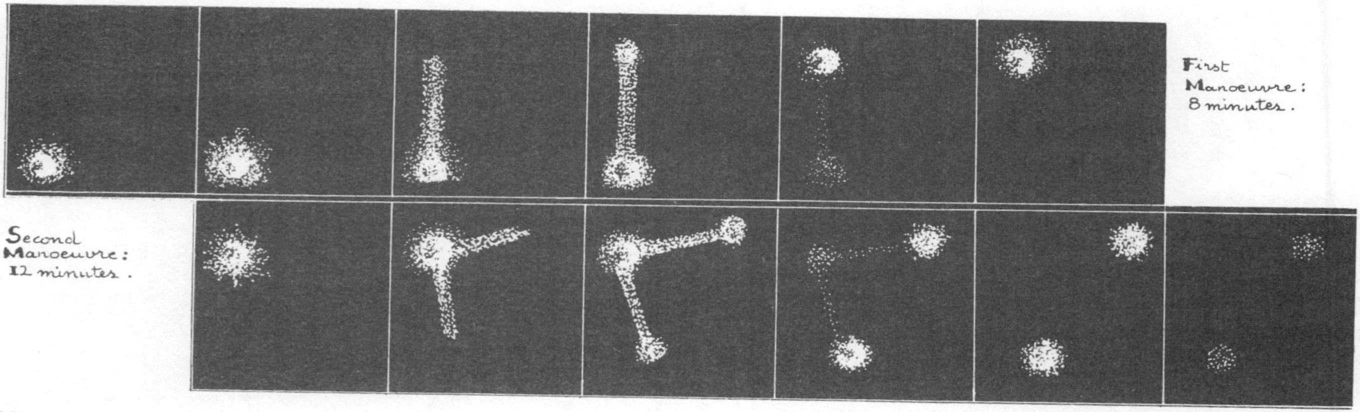

Spherical light,	increasing in brilliance,	projecting beam(s) while diffusing,	regaining form while secondary sphere(s) appear(s) at end of beam(s).	Secondaries increasing in size while primaries and beams vanish.	Resulting spheres,	vanishing. ~

bay as darkness set in, bringing my journal up-to-date with the story of my final swim, taken earlier the same evening.

I had just written:

> As for my final question, apart from thundery skies, straight out of *"Close Encounters,"* we haven't, as yet, received any impression of U.F.O.'s, or of the dolphins' possible relationship with them —

I got four lines into describing my suicidal attempt to meet the dolphins face to face when I was interrupted by a cry from my companion. It was 9:05 p.m.

A very bright light had appeared offshore, about 30° above the horizon. We ran to fetch the binoculars and a small telescope we'd noticed earlier and, returning to the balcony, could quite clearly observe the light increasing in intensity.

As we watched, it projected a vertical beam some distance above it, decreasing in brightness as a smaller ball of light slowly appeared at the end of the beam. The first light disappeared when the second light had grown to the same size as the original.

It was all quite astonishing! We spent a few minutes going through the possibilities, rational beasts that we are!

It was a dark cloudy night with no stars visible. The waning moon was nowhere to be seen and the light was in the wrong place for Venus, even if we could have seen it. Ball lightning, we supposed, could have been an explanation — neither of us had ever seen ball lightning before, but it features in much U.F.O. debunking, so we took it into consideration. But lightning that stayed in the same spot for almost 20 minutes . . . !

Thunderstorms were in the air, they'd been lighting up the Westerly sky since it had grown dark but, in contrast to those flashes, the constancy of the balls of light, with their regular movement, suggested little to do with natural phenomena!

Apart from the manoeuvering, illustrated by my companion, the whole configuration remained in the same place for the entire period of the sighting. Scarcely likely it was some coastguard flare — unless they'd invented skyhooks.

We could also see clearly the reflection of the glare on the underside of the clouds. To this day I have no satisfactory explanation for the lights and am left with the rather startling thought they had appeared in direct response to my earlier paragraph, written only two minutes before they first made themselves known.

After this curious interruption I returned to my Journal entry.

— It was around 5:00 p.m.

A single large dolphin breached opposite the condominium, about thirty feet out. I waved to him and had the sense of being 'seen.' Dolphins, we were discovering, have excellent sight out of the water and literally jump to see.

It was all the provocation I needed. Perhaps this would be the moment they'd give me for the opportunity to swim with them.

The beach was deserted by the time I got down there, the darkened skies and intermittent rain lay sombre over the gathering waves. I threw my clothes down next to a small piece of litter, a toy dinosaur, and waded into the tepid water.

I had just passed the first sandbar when I saw a dark form slide under me. I knew the dolphins might be there but they still wouldn't make themselves obvious. At the next sandbar I got my breath back and then jumped up and down to see if there was any sight of them. Nothing . . . For some reason I'd even rejected the shadow movement under me as a sign of their presence. I was in a very curious state of mind!

By this time the water had become choppy and lightning had started to crackle down near to where I was swimming. I had reached another of those turning points.

All the extraordinary incidents of the past week came mounting in like the waves all around me. It was now or never and I desperately wanted some kind of objective verification of all the telepathic impressions. Faith and belief are all very well but something inside me needed to know for certain I wasn't fooling myself with a lot of preconceptions. Mad though it sounds now, I decided to take the Rastaman's example in earnest and swim out to sea as far as I could go before my arms and legs seized up. I'd come full circle on this issue and By God! If the dolphins wanted to 'rescue' me and it was the only terms on which they'd let me close to them, then rescue me they could.

I set out into the storm. After about twenty-five minutes of swimming, the wind blowing globs of seawater down my throat and cramp starting to get a grip on my left leg, I recall thinking "it's now or never, lads . . ."

Nothing!

The bastards, I thought. They're going to let me drown! Well, I deserved it — taking on this ridiculous macho swim hadn't fooled anybody down there. I was 'in contact' with them, that I knew, and it was starting to have a curious effect on me. Try as I might, I couldn't panic. It was the same as last time. They must have entrained my psyche in some way so I couldn't even feel the fear I had every reason to be inundated with. The condominium looked like a matchbox on the swaying

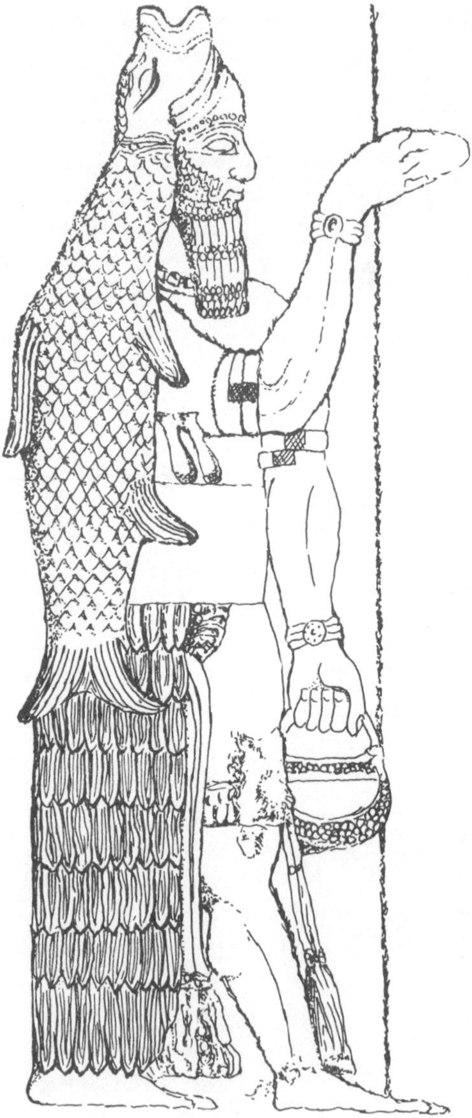

horizon yet I couldn't take my absurd situation seriously.

I started laughing uncontrollably, knowing the tricksters had outsmarted me again. I'd become obsessed with swimming up close to them, riding their dorsals like the sculpture of the boy on the dolphin. This fixation had percolated up from my deeper mind until I was attempting feats beyond my reasonable level of expectation.

It was in those moments I had to acknowledge I was meeting both dolphins and all intelligent denizens of the Inner Worlds entirely on their terms. I welcomed their using me in whatever experiments they wished and knew too, in that time of extreme stress, they wished me no harm.

Knowing this, and feeling a new strength welling up within me, I turned back to the shore. The sky was brightening and the storm had moved over. It was a long but surprisingly exhilarating swim. Twice my adrenalin rushed as I felt a very fast but gentle stroking across the back of my legs . . .

Arriving back over the sandbars, I waded the last few yards. As I did so I had the most powerful memory — I can only call it that but I experienced it as a sort of superimposition, an overshadowing — Of OANNES, the mythological creature, half-man and half-fish who walked out of Babylonian waters five millennia ago. Where the impression came from I have no idea, but I felt the great fish head straddling mine and the pendulous, slow-motion swinging of the huge, bipedal body . . . the strangeness of the land and the tasks ahead . . .

On arriving back in the condominium I was greeted by my companion, happily smiling that I'd had my way. "Had I seen the dolphins swimming and leaping all around me?" she asked.

"Whaaat? All around! You must be kidding — I saw nothing . . ."

The realization was slow in dawning on me. Apparently they'd accompanied me every yard of the swim. My companion had been able to see their dark forms gliding under and circling me. If I turned towards the sea, one would breach behind me; if I looked back at the shore, they'd leap again exactly in the direction I was not looking. And after making my swim and while I was clambering back up the beach, she'd watched them making one last tour of inspection of where I'd been, before taking off for the open sea.

It was an unmistakable signal and one not wasted on either of us. We'd gotten as close to these wonderful creatures as they would allow us and there was nothing for it but to return to New York and digest all the impressions we had received.

Neither of us guessed the next stage of this initiatory journey would come so fast, but it did and we were both grateful for the lessons of casual ease which we'd picked up from swimming with the dolphins.

CHAPTER THREE

Labor Day of 1981 fell on an overcast Monday. It never seems to be the easiest holiday of the year, with everybody bored and having to cope with those insistent, disconcerting thoughts that flood in when hands are idle. This inevitably creates psychic disturbances in the 'thoughtsphere,' as definable to the sensitive person as are shifts of weather in the physical world.

The ebbs and flows in emotional wellbeing are particularly powerful in large cities and those of us who choose to live in them have to work additionally hard to keep on top of the general malaise.

So it was with that Labor Day Monday and when we rose late it was to find ourselves locked in one angry disagreement after another. Letting it out and discharging the pent-up emotion invariably helps lift the weight and clears the sensibilities as well as detaching the individual psyche from the pervasive negative atmosphere.

The sun started to appear as our moods brightened and we took the opportunity to go out and photograph the graffiti we had sprayed the night before.

Although I've long had respect for the courage and artistic enterprise of the better among the New York Subway graffiti artists, I had never before the inclination to turn my hand to it.

However, when the local authorities decided to rip up the concrete park benches in the small square only a block from where we lived in Manhattan, an opportunity for disposable graffiti presented itself. My conscience was assuaged by knowing the large, flat slabs would be removed in a day or two and they provided a perfect surface for a bit of inner city artwork. The fact that the site was Bennett Park, the highest natural elevation in Manhattan Island, added a geomantic consistency which made our midnight foray all the more personally significant.

On our way out that Monday we chanced to meet a neighborhood friend on our doorstep so all three of us went out to the Park together.

The graffiti looked quite splendid in gold and white, and pleasantly humorous too. We found an old man already photographing it and chuckling at the erudition of the kids these days . . . We didn't disabuse him, leaving him wondering to himself what on earth THE LIZARD HAS LANDED might have meant.

When it came our time to photograph the painting, my companion, checking out light levels, chanced to look up. She exclaimed in surprise and we followed her gaze to see a sizable, moving disc travelling low and steadily across the grey sky. It was a blueish/green color, and about the same size as an American 1 cent piece held at arm's length. It was well below the cloud level and quite distinct.

The three of us watched in a state of absorbed fascination as it transitted the sky in about thirty-five seconds, disappearing over the treeline in a North Westerly direction towards the Bronx. It was difficult to judge its size because there was little to which to relate it, but I would estimate it to be about 20 - 30 feet in diameter and travelling around two to three thousand feet up. Its motion was steady and regular and it had the look of being powered, unlike a balloon or a meteorological kite. As we watched, it fluttered briefly without changing speed or direction, as if it was tipping its wings. It was at that point too, my companion managed to get a photograph of it with her Brownie Box camera.

We had not talked in any detail about our recent sighting of the lights in Florida, sensing in them a mystery yet to be revealed. This flying disc, with its concrete reality, and seen at 1 p.m. on a clear Monday afternoon, however was quite different. I recall thinking as it passed overhead that if I'd ever have asked for a bona fide UFO then I could expect nothing more real and down-to-earth than this remarkable apparition.

I had finally learned through the Florida episode one very important factor in approaching manifestations of the Inner World. Act casual! However excited or elated one might be inside, to maintain the rhythms of the experience, it's beneficial to stay cool.

This learning was to be of great value in the events that then took place.

Within moments of the craft's disappearance over the treeline, a young boy who had been standing, unnoticed by us, a few feet away and also watching the departure of the UFO, turned to us and announced in an off-hand manner that it was merely a "starcar."

I felt a ripple of primeval fear go through me and the hairs on the back of my neck stood up. There was a certainty about the child's comment, a familiarity and lack of protest which added conviction to his announcement. Besides, I'd looked after an eleven-year-old boy for the previous few years and I knew 'starcar' was not in the lexicon of current science fiction comics, films or television.

"Sit down here young man," I said, "and tell us what a starcar is."

"It's a single-person craft used generally by tourists and explorers," he replied in a thick, what I took to be Middle-European accent, "and it is connected to a mother ship by its central drive mechanism."

It came out very fluidly, without any pause for thought, and I had the oddest feeling something very profound was transpiring. Perhaps he could have concocted the idea in the way of lonely, outcast children but there was something in his manner . . .

He was somewhere between seven and ten years old, chubby, and had about him an androgenous male/female quality that might have been garnered from long days spent indoors among musty, overstuffed furniture. The tradition of the anti-macassar dies hard in those old German-Jewish enclaves. But the lad seemed to be onto something, so I asked him to continue telling us about his starcar.

"Well, they certainly can travel between planets," he continued, eyes glittering at being taken seriously, "but they invariably relate back to the mother ship. You see both the larger ships and the smaller starcars use a Mani Particle Beam propulsion system that is picked up by a central unit within them." He was warming to the subject. "This unit establishes a particle beam rapport instantaneously with a terminal within the giant star PHINSOUSE . . ."

"How do you spell that?" I asked, scrabbling for the inside of a cigarette pack. He patiently waited and then, in that manner of children indulging the slower ways of adults, gave it to me, a letter at a time. Smart-ass kid, I thought, but since I had pencil at hand, I'll go the whole way.

"Does it have a Universe number?" I asked.

"Local or General?" his eyes were amused.

"Local first."

"79562183" "And General, 1765333177082"

When I'd written the numbers down I asked him to repeat them back to me, which he did correctly. The encounter was starting to take on a new dimension. How many kids, I wondered, could have retained twenty one digits and then repeated them back without faltering?

"The star system Phinsouse is inhabited by a very old race of non-space faring beings. It is for this reason it has been chosen as the center of space activity. There, every inhabited planet has a separate chamber, which represents the race concerned and the progress it has made."

"Where is Phinsouse?" I asked.

"It's in the center of the Andromeda Galaxy," he replied without a pause, "the most populated galaxy in the Universe, with one million races living on three billion planets."

Fortunately I'd been prepared for these massive figures since I was currently reading a large and impressive tome called *The Urantia Book.** This book, with its encompassing Universe cosmology will reappear from time to time through the unfolding of this tale, as its impact and meaning made itself felt on me and the journey through which I was being led.

Suffice it to say here that it purports to be a description of the inhabited Universe, the manner of its civilization and spiritual/scientific knowledge. It also speaks of a vast and densely populated Universe with billions upon billions of individuals living on countless worlds throughout "pervaded space."

I was anxious as to the form of those races our

* Publ. by *The Urantia Foundation:* Chicago, 1955

young friend was mentioning so casually.

"Most of them are humanoid. Some races . . ." and he gave a name which I didn't catch, and not wanting to stop the flow, ". . . are as big as . . .", he looked around for a comparison, ". . . as houses — and they constantly change form. Those are in the minority though; in the main the humanoid is the most popular configuration."

What unites all these different races? I wondered.

"All entities with very few exceptions" and he emphasized **exceptions** "have feelings. In fact it is feelings that all sentient beings have in common."

I recall having an inner sense of great joy when he said this.

"It is precisely these feelings," he went on, "that form the basis of the Mani Particle Communication System. Through this, matter can be transformed into three-dimensional images and projected over vast distances."

I asked him what he meant by "re-constituted into 3-D images" and he corrected me with some insistence.

"Not re-constituted; transformed!" Very certain of himself. So I put my hand on his chest, which certainly felt real enough, and asked him whether he was a three-dimensional image.

He looked at me for a long moment and giggled as if caught off guard.

"No," he replied. "I'm not." We all laughed together.

"Tell us more about the Mani Communication System," asked my companion.

"Well, it's still in an experimental stage," he answered, happy to be consulted. "Images can be projected, but no one has yet been able to successfully transform matter itself and project that." Research, he told us, had been going ahead very slowly because it was so dangerous.

"Some people come back . . ." he rolled his eyes in compassion, ". . . a little damaged." He was quite obviously very distressed by this and my companion and I caught each other's eye and both involuntarily shuddered at what "damaged" might have meant.

I asked him about the political systems used out there, to move away from a subject which so evidently caused him pain.

"The most general one used is a one-person, one-vote democracy called Anthesis. Kings and Emperors do exist, but only in small backward places — in about twenty races — and are considered rather primitive."

"Is war a factor out there?" I asked.

"Yes, there are a few warlike races but they don't get very far from their home planets because the lines of supply become so attenuated. War takes a lot of energy and expense," he smiled again to himself, "and in those cases, the races involved and the physical spaces they occupy, are simply avoided by everybody else."

There did not seem to be any intervention policy. The belligerent races were quite simply allowed to learn for themselves.

Let us pause for a moment and take stock of this thoroughly weird situation. There were the three of us; Stephen our neighbour, my companion and I, sitting on the flanks of the great rock which surfaces briefly at this, Manhattan's highest point. Manhattan, you will remember, is built on the back of this single, gigantic rock dome reputed to be the top of a vast underground cavern. About the cavern we do not know for certain, but the massive granite outcropping certainly does appear and re-appear as you journey up from the financial district, baring its haunches in Central Park and then again in the

Cemetery at 157th Street and Broadway. It re-appears in Bennett Park before dropping away under the Cloisters and Fort Tryon Park and finally falling into the waters of the North River at Spuyten Duyvil.

It is because of the solidity of this great rock that the engineers and architects of the last two centuries have been able to erect their soaring skyscrapers without danger of subsidence. But, as anyone who knows their rocks will tell you, a large solid mass like this represents a single homogenous unit with untold implications for those interested in mineral intelligence systems.

Perhaps our graffiti had acted as some sort of sign to the denizens of the flying disc. Perhaps this small androgenous boy to whom we were talking had inadvertently become their telepathic mouthpiece.

It was every bit as surreal as my ridiculous last swim with the dolphins in Florida, yet normal, grey, blank New York honked and hooted around me as if normal reality just continued unchanged by these encompassing revelations.

And could the boy simply have been making it all up? In his loneliness and isolation, perhaps a Rabbi for a father, filling his head with all these fantasies of a teeming Universe . . .

But this was elegant stuff! Mind-boggling concepts, and I knew from experience how few people can even accept the true implications of an inhabited Universe with all its implied impact on our terra-chauvinistic worldviews. Good Lord, even the most advanced scientific thinkers are only now starting to struggle with the idea that we may not be alone, a happy coincidence of elemental chemistry brewed from the primal ooze.

The kid hadn't finished with us yet however, and the tone of his discourse had changed. He was standing now, walking backwards and forwards as if lecturing a small group of students.

"There is another political system," and he paused theatrically, making sure we were listening to him, "which is used sometimes in the case of planetary cultures which do not recognize they exist within a Universal community. It doesn't have a name but it functions in the following manner."

He drew himself up straight, bending over every once in a while to emphasize a word by peering deeply into one or another of our eyes.

"All civilizations have problems, some big and some small." This from an eight-year-old, I recall thinking to myself as he twisted the plastic disc he'd been holding and stabbed the air with it for emphasis.

"And there are those people who take on the major issues, ponder them and give them consideration," he paused to make sure we understood, "and it is they who become the vote carriers. They become the hidden leaders of their races and are permitted higher roles to play."

That makes sense, I thought, in a benign and democratic Universe. Obviously decisions are going to have to be taken which influence planets that may be wholly unaware of the larger community. Of necessity, there are going to be people called on to represent their planetary brothers and sisters. It wouldn't be fair otherwise.

The last subject he covered with remarkable passion and a genuine sense of wonderment. He said there had been considerably heightened interest in this planet since the Second World War but that no detailed records had been taken for the last ten thousand years.

"The last film we saw about this planet was 10,000 years ago," and then in his strangely clipped European accent and his sage young/old eyes, he pulled himself up to his full, pudgy height and announced with overwhelming authority, "Caveman to Civilization overnight . . . Caveman to Civilization overnight!"

Throughout the interchange the boy had been fiddling with a plastic disc. We had asked to look at it but he'd been reluctant to let it out of his hands. He said it gave him energy for physical things since he didn't have much of "that kind of energy." It looked to me like one of those flexible, clear plastic fish-eye lenses which you see in the windows of novelty shops, but he wanted to prove his point about energy.

He put down the disc carefully on the rock beside us and ran around in a thirty-five foot wheeling circle. The first two times he plodded along without the disc and was clearly ailing by the end. Then he picked up the disc again and circled another five times — it was difficult to stop him — and although he appeared to be in a mild hypnotic trance he was undoubtedly in a better condition when he'd finished than when he completed the earlier two cycles.

Our conversation ended after about twenty minutes. I'd asked his name three times but each time he'd evaded it and continued with some new tidbit of fascinating information. I didn't want to push him since he'd probably been told by his parents not to talk to strangers.

Neither would he allow me to buy him an ice cream and continue talking — he said he had to get back home. Although intensely curious we maintained our casual approach and made no attempt to follow him. Neither have any of us seen him in or around the Park in the years since the encounter.

When the photograph of the craft was developed, there was indeed a small spot in the middle of the frame, though smaller than the object appeared to us. This may have been simply the limitations of a Brownie Box but we did not dismiss the possibility the disc may have been enlarged by our perception of it. These matters can be very subjective.

Well, **could** he have fabricated the whole story? Obviously, given the limitless potential of the human imagination, the child might have been able to provide a coherent Universe view. Interestingly enough, I was at the same time attempting to sort out a similar issue in relation to *The Urantia Book.* This 2000 page document, as any will agree who have seen it, maintains it is transmitted from the celestial and angelic beings who create and watch over the inhabited Universes. Its long and detailed descriptions of life on other spheres apparently "came through" the mouth of a sleeping man in the first quarter of the Twentieth Century and raises the identical issue posed by our encounter with the boy.

Could both discourses be the product of an enormously fertile, but thoroughly human, imagination?

In both cases we have nothing to which we can compare the content. It either rings true to us or it doesn't, and I am firmly of the opinion we humans have an internal faculty which can discern the higher Truth of such matters.

As for the boy, his delivery was certainly not premeditated. It wasn't a spiel he might have thought up in his long hours alone because it was very much a dialogue with our questions going well beyond the scope of easily anticipated areas.

His manner was forceful and way beyond his years,

and precise to the point of vehemently correcting me when I'd used the word "re-constituted" rather than "transformed."

He made good consecutive sense in so far as any of our knowledge of science goes. The piece about war is manifestly coherent, as both Napoleon and Hitler discovered in their Russian campaigns. It is also not a matter which springs immediately to mind in a generation reared on the evil empires of Darth Vader or the Klingons.

My companion reports too that the child answered a puzzling aspect of our Florida sighting. She'd wondered if the lights were actually demonstrating something, a physical principle or a shape or form. When the boy had started telling us about the Mani Particle Beam as a method for projecting images, the lights started to make more sense. The secondary light source had appeared, in each case it had occurred, as a projection of the primary source although at the time we hadn't understood the significance.

"Caveman to civilization overnight" and "The last film we saw on this planet was 10,000 years ago," were undoubtedly his two most memorable statements and, while he delivered them, we all noticed a clear change in the tone of his voice and manner. He took on a profoundly deeper and more aged persona. While most of his dialogue was spoken with an evident and intense wish to communicate, he made those two observations seemingly to himself.

Whether there was in fact a direct connection between meeting the boy and seeing the object in the sky we can only speculate. It certainly felt that way! There were just too many coincidences within and surrounding the whole event to avoid seeing an associational line between the dolphins, the Florida sighting and now this unfolding drama of the UFO and our small, alien mouthpiece.

CHAPTER FOUR

Up to this point, I had focused on cetaceans in their wild state, assuming from the few film clips I'd seen of oceanariums that captivity certainly seems to dull their light. My distaste for zoos kept me away from observing these creatures in confinement: I, like many others, had wondered why species as apparently intelligent as these would ever allow themselves to be caught and paraded with such indignity. If it turns out this was the only way **they** could reach out and contact **us**, then it would be a sorry comment indeed on the state of our own sentience!

To be fair to our investigation however, a visit to one of these prisons had to be undertaken and, as these things happen, we chose predictably an example at a time of immediate crisis.

We arrived at New York Aquarium in Coney Island on an overcast August afternoon. On entering, the first sight striking our senses was the idiosyncratic vision of three white Beluga Whales, literally at eye-level. The moment was one of some poignancy, the accompanying complex of emotions profound enough to bring on a short attack of cognitive dissonance. It ranged from an underlying sense of fellowship and swung over to the sickening horror of seeing these inexpressibly languid, beautiful creatures trapped in glass cells only feet larger than themselves.

Closer to the Belugas, I had an immediate sense of contact, of being seen and registered. I had no idea of how much they could 'read.' My companion and I stood at the rail, two among the fifteen or so people who came and went over a ten-minute stretch, and we soon became a point of interest for the three Belugas.

Tears were starting to roll down my face by this time, and I felt myself enter the same sort of rapture which marked my contact with the dolphins in Florida. In this strong emotional state, and with a feeling of rapport with the whales, I tried some simple experiments. If I formed a straight-forward question in my mind like, "if you can hear this, nod your head," or by asking one of them to swim in a circle, and repeating the phrase clearly in my imagination a few times, I found they invariably complied. Their swimming in a circle set my adrenalin pumping; although sinuously slow, it was so direct and pointed, with even a significant glance from the Beluga to me on completion of the turn, that I moved from my heart into a stream of ideas. Possibly the concepts were my own, stimulated by cosmic consciousness, the metanoiac hit of telepathic contact with the whale, but the stronger sense was of a transmission, an open line of communication between its consciousness and mine.

The vision was brief, powerful and culminated in the perception that the whales and dolphins have a definite plan of action which is benevolent and profoundly intentioned towards the mutual benefit of all. Layered over this came intimations of a great joke! A plan involving some sort of Divine humor, and one in which everyone will get exactly what they want. Then the idea stream faded.

Sometime during the ensuing half-hour spent with the Belugas we overheard the staff were having problems with a newborn Beluga calf. The mother had refused to feed it and the calf, although force-fed by the aquarium staff, was in a poor state and ailing fast. There was an accompanying feeling of mild panic since this was one of the few Belugas born in captivity.

We offered to help although we had little idea of what we could do. The young woman went away to convey the message to those with mother and calf.

Falling back into the empathic rapport with the three Belugas while we waited, both my companion and I received an impression of complete dejection from them. We found out later the father, who had attempted to kill the baby and was thus separated, was among the three. The impression continued with a sense of the pointlessness of the birth; that these cramped surroundings were no sort of environment into which to bring a baby. The emotions were so raw I was sobbing again by the time the assistant returned with the news that "everything was being taken care of."

"The professionals are on the case," she assured us firmly.

There was little we could do outside of barging through the red-tape, and this could have well been counter to the Beluga's personal interests. We begged to be allowed in, citing some of our recent interactions: There just might be **something** we could have done, if only to have telepathically projected our condolences to the mother. But, after coming and going one more time, "thanks but no thanks" was the answer.

The whole incident had been suffused with a profound awe. We had no doubt we were watching an ancient tragedy being played out by a tired, but infinitely patient species in its desire for contact and communication.

We passed the time before the Dolphin Show standing looking over the railing into the dolphin enclosure. Beyond it, separated by a low barrier was the other pool, in which swam the mother and baby Beluga. While we watched, the three dolphins spent almost all their time up at the barrier, as near as possible to the mother and calf. Their intentness to be close to the Belugas becomes significant within the context of our next experiments.

I had spent some time watching the sea-lions by myself before rejoining my companion, who stayed by the dolphin pool. This is what she relayed to me.

Having had strong mental contact with the Belugas, she was interested to find out if the same form of communication worked with the dolphins. She 'thought' for one of them to swim the intervening 35 feet, and attempted to project the thought to the dolphins. No response. She tried again. Still nothing.

Not wishing to push her luck and knowing there was a delicate situation afoot, she stood quietly watching the dolphins at the far side of the pool. She related how at that point, her thoughts had turned towards me, thinking of me sweetly and with all the pleasure of being in love. The thoughts became more playful and sensual as she dwelt on them. Within moments, one of the dolphins swam straight over to her. She came out of her reverie in surprise, realizing the dolphin had responded directly to the strongly felt mentation. She tried it with more consciousness and again it worked. The third time she mixed the feelings with a projected thought for the dolphin to swim over and again, there it was, with no question it had seen and was looking right up at her.

It was soon time for the Dolphin Show and other people gathered around the entrance waiting to be let in. The dolphins showed no sign of interest, preferring to stay up the other end until their trainer appeared and the excitement was due to start.

On being allowed in I headed off to reserve some good seats while my companion remained at the poolside to see if she could get a close-up photo of the dolphin swimming over to her. The earlier mood had dissipated and the dolphins weren't responding to projected thoughts, so she tried another tack. Being an artist and having therefore developed an unusual level of visual acuity, she decided to see whether the dolphins would respond to an image. She pictured a dolphin swimming over to her as lucidly as she could in her imagination. Just as promptly as before, a dolphin sped over to her — and there'd been no other action in the tank throughout this period. Her photos show him on the way over and another of him returning, throwing a curiously direct look back at her over his fin.

Never having seen a full dolphin performance before, I imagine the range of tricks they went through when the show finally started was pretty nominal. Not unreasonably, the trainer must have had a lot on her mind and wanted to get the whole affair finished as soon as possible. But even this did not blur the beauty of watching these extraordinary animals circling the small pool at about 12 knots and leaping through impossibly tight situations in perfect synchronization, high out of the water.

During their performance neither of us received any particularly strong feelings from the dolphins beyond a natural ecstasy at seeing them so close. Dolphins definitely do 'notice' individuals in the crowd, especially happy people, children and lovers. We'd heard this anecdotally from other trainers. Although somewhat conjectural, if this is true, it could well substantiate our growing conviction that captive dolphins are attempting to make themselves 'felt' to receptive people, especially children, and are indeed using Dolphinariums for just this purpose.

So, what had we learned from this deeply ambivalent encounter?

In the case of both Belugas and dolphins, we confirmed to our own satisfaction that thoughts appear to be communicable across the inter-species barrier, when associated with strong and genuine emotions. In addition, we had some preliminary intimations of a potential visual channel of communication between dolphins and humans. Both these areas of investigation could prove very fruitful and we decided to pursue them again when we were able to spend more time up close to either captive or wild dolphins.

Shortly after returning from Florida in late August of 1981, I started corresponding with Colorado-based astrophysicist Gerrit Verschuur on the recommendation of the Human/Dolphin Foundation. He was one of the few people they knew who was actively investigating the telepathic factor in dolphins.

His letters were articulate and psychologically interesting, underlining the more metaphorical and mythological aspects of the search for alien intelligence. He had spent time swimming with John Lilly's dolphins and had written up two excellent proposals for a technical study of radio and electromagnetic waves from the viewpoint of dolphin communication. Among the many points mentioned, he urged us to keep a very close watch on our dreams, as he had noticed a trend towards significant dreaming among many of us who have become involved with cetaceans.

The dolphins had certainly seeped into my dreamworld; this became very evident through the long, hard winter. Either they in person, as it were, their artifacts or, in some cases, connecting links in the story they appear to be unfolding for us. These were seldom the usual daily re-run type of dream, but clear and incisive ones with a sense of coming from 'somewhere else.'

Six days after the second UFO sighting over Bennett Park, I had a dream which I strongly associated with the dolphins, although none actually appeared in it. I noted at the time:

(Dream Fragment)

I am being shown a small stack of what appear to be long crystalline structures. They are neatly piled and, as I watch, I can discern two different types. On closer examination they turn out to be quartz crystals about two and a quarter inches long with a hexagonal cross-section of about three eighths of an inch. The ones at the bottom of the stack appear to be plain crystals, but the ones at the top — I would estimate about twenty of them — have a very fine gold covering that shimmers in the light. The hexagonal surfaces of both gold and crystal are precisely tooled and highly polished, as are their end-sections.
I awoke with the strong feeling I have been told the crystals have to do with a new form of energy package.

(End of fragment)

I decided to duplicate a single crystal as near as possible to one of those seen in the dream. My first call to a lapidary produced immediate results. Quartz crystals of the dimensions I needed are extremely rare — usually they are thicker or less regular and have to be cut down — and I wanted to avoid tampering with the natural form itself, but, yes, there was one he could let me have at a remarkably reasonable price. Quartz crystals, he told us, had tripled in value since they started being used in computers.

I needed a second lapidary to cut the crystal and, coincidence again, one surfaced able to do it of all days, on a Jewish holiday. He must have been the only Gentile lapidary in New York City! Having cut and polished it, my companion was able to complete the gilding using a gold-leaf process she unearthed from the writings of Renaissance artist Cenneno Cennini.

It was all coming together and now, for whatever it was worth, we had a very beautiful, piezo-electrical device, the purpose of which we had little notion. Perhaps most significant to us, however, was the extraordinary ease with which the whole operation carried itself out. Very much as if the 'someone out there,' whoever or whatever had perhaps implanted the dream, wanted it made and had cleared the way by orchestrating the run of synchronies.

It is worth noting at this point that throughout our explorations of dolphin intelligence we found ourselves pondering the recurrent appearances of hexagonal forms, invariably associated with unusual events. In drawing up both the sand dollar and the first, lights-in-the-sky UFO sighting, my companion found the same hexagonal shape subsuming both forms. I wondered at the time whether we had stumbled onto a cypher or code of some sort. A sign of the involvement of higher powers, a trademark, possibly an imprimatur.

The dolphins remained very much with us in our minds throughout the ensuing months in New York. I spent time recapping on the various incidents during my life in which I had made some sort of contact with cetaceans. Like many landlubbers, these moments of communication have been mainly through books and films. Joan McIntyre's *Mind in the Waters*, a magnificent paeon to cetacean intelligence from the scientific to the poetic; the disturbing *Day of the Dolphin;* the inaccurate but haunting movie *Orca;* and a number of other occasions when dolphins or whales were shown on television, inevitably contrived to produce in me a quality of rapturous awareness I have since come to associate with a radically heightened consciousness.

All very delightful, I thought, nevertheless I would find myself invariably deposited in an elevated state of consciousness by some factor these entities possess and about which I had little or no understanding. Even though wonderfully uplifting, the impact of their presence rendered me incapable of anything but an ego-loss reaction. I consistently found myself in tears of rapture, heaving sobs of joy and sadness; my heart prised open in some way by contact with cetacean consciousness, and yet reduced to inactivity by the sheer force of what I can only imagine is the different, and perhaps higher, vibrational frequency of the dolphin/whale intelligence system.

I recalled too, a time when I had played Paul Winter's *Common Ground* album, one track of which features a skillful integration of music and whalesong, while peaking in an LSD session. After empathically bonding with a whale and following her progress through the oceans, I found myself, in one vast change of scene, prone on the slick, metal deck of a Russian whaling factory. I lay shudderingly still, panicked for an eternity, before sliding slowly down into a wall of circular knives which stripped and cut my great body into thin slivers.

Rethinking these incidents in the light of our Florida experiences made me realize that if I am substantially correct in my perception of a high level of cetacean consciousness, then interaction with these creatures may well be creating some form of rhythm entrainment that automatically raises the consciousness of the human participant. Obviously, sensitivity and practice in the area of altered states will greatly increase the level of control and understanding of the rapid accelerations.

I was ruminating on these matters one evening in late November, doodling meditatively in my journal, when I clicked back into focus to see that I had written, without any conscious realization, the following:

> *"This planet, at this time, is witnessing the birth of a Conscious Universe. This is all but incomprehensible to you because your neurological capabilities in a normal waking state cannot make the necessary quantum leap. You can sense its intimations. Feel its shadow. There are places you can pass through, experience life in the Conscious Universe and report back . . ."*

I hung suspended, still entranced, watching my pen bob and flutter. The passage continues:

> *". . . so hard, so hard to retain the fire, the passion for change. The recognition of a perfect Universe depletes our vitality . . . saps our will. There is nothing left to do, or worse, everything we turn to do, or be, becomes pointless. Because it is all becoming anyhow . . . But this has not to be so. The perfection rests on each part, is contained in each part, to such a degree that every movement of the part is felt directly by every other."*

I realized, through a soft, white haze, that there was something of the dolphins in this. Had I fallen entranced and allowed them to write through me? Was it my superconscious at work, writing from some other level of awareness? Was it a voice from the angelic realms? I caught myself before I fell into a level of mental rationalization I knew would block the communication, if that is what it was. I felt through the cetacean's thirty million years of consciousness, of self-consciousness even! What **could** that span of time mean to us?

The writing broke through again but this time I felt it intermingle with my own feelings . . .

> *"It is enough simply to be? To be for whom? For God, of course, but why?*
> *And, if we choose not to be, does that alter the picture one jot?"*

Rationalization and self-consciousness had taken their toll. I no longer felt in contact with an external force. My mind turned to wondering whether dolphins and whales might have climbed up an evolutionary ladder of consciousness, as have humans in the manner suggested by R.M. Bucke in his seminal book *Cosmic Consciousness.* These changes seem borne out by human development over our paltry million years of growth — might it also be a pattern shared by other sentient beings on the planet?

Whatever could it mean to have carried consciousness for all those millions of years? At this point, a feeling of profound ennui washed over me. Endless cycles of life and death. The passage of sun, moon and season. Interminable lives of placid playfulness and languid procrastination . . .

As the ennui faded, I recalled the feelings we had with our wild dolphins in Florida. I saw it too in other cases of individual dolphins that have, on their own initiative, made contact with researchers; the sense we are being approached by the leading edge of the cetacean community.

If they can be said to have possessed consciousness for so long and to have carved for themselves such a secure ecological niche, then they probably have an appropriately conservative social order. If indeed this is so, then it may go towards explaining why individual dolphins choose to break away from their schools and take the time and considerable trouble to develop relationships with humans sensitive to their situation. Not dissimilar to social analyst Philip Slater's perception of human societies, in which a host organism, a cultural unit, extrudes a prophet, who then ventures into new territory and reports back, thus increasing the boundaries of the entire culture, these solo dolphins may be aiding the overall cetacean community inch forward in its real understanding of us.

In a coherent and benign Universe, it must be inevitably true that our two species are making a profound new contact at exactly the relevant point in each of our developments. While it is somewhat easier to see what social and material gifts the dolphins have for us, it is also important to appreciate that we hold many keys for them. Not only do we have the opposable digits, for instance, to get their bodies into space, but we inadvertently present them with the exquisite challenge of comprehending an entirely different form of sentience.

And, if by upstepping our human consciousness, they are guiding us towards living like gods, what is that supposed to mean? What are the implications of living in an on-going state of cosmic consciousness? What is a greater intelligence?

What can it grasp that we in a usual state cannot? Does it have limits and does it have its pitfalls? Is there such a thing as partial omniscience? What challenges would be left to an entity with this level of grasp of reality?

With these ponderings I realized I was opening up a new line of questions not dissimilar to those I had considered before we left for Florida. I wondered what lay in store for us in the answering of them.

In early December of that year, I had another startlingly clear dream. I noted at the time:

> *"I see the dolphins swirling all around me. The hexagonal shape becomes very important. The scene shifts to five people, and myself makes six, viewed from above and standing in the form of a large hexagon. It is a power configuration of some sort, perhaps a telepathic receiver — I can't tell. I see superimposed on this the first of the two sightings, with a hexagon drawn white on black over the positions of the lights."*
>
> (End of fragment)

Most significantly, this dream coupled the dolphins/hexagon with a suggestion of both energy and telepathic reception — factors that are turning out to be integral with this unfolding journey.

Somewhat later, and nearer Christmas, a friend gave me a copy of Ted Mooney's remarkable first novel, *Easy Travel to Other Planets.* The book essentially deals with the dolphin/human interface, interweaving in punk-existentialist terms the destinies of a small and sophisticated group of New Yorkers as they fall irrevocably into the telepathic twilight zone of a dolphin's spell. It is a superbly poignant novel which covers the tones and moods of dolphin/human exchanges with such a complex perception that it is hard to believe the book is not channeled from the same telepathic domain it describes so beautifully.

Mooney portrays the dolphins as dreamers and lovers, capable of handling immense amounts of data simultaneously. A race of telepathic poets, aroused by their concern for the environment into making closer contact with humans, and who, Ted Mooney implies, may already be permeating our sense of reality with their vast and timeless presence.

Having swum with the dolphins, and having spent considerable time pondering the more prosaic aspects of cetacean intelligence, Ted Mooney's novel triggered in me the same feeling I experienced in Florida while looking through the Audubon book. I had the distinct sense the dolphins had somehow 'arranged' for me to discover, through a novel, granted, more details of their lives. In fact, the accelerated synchronies the author puts into a fictionalized context have been starting to take place in my life since my interest in dolphins took me to Florida.

I am tentatively suggesting at this point, therefore, that there may be some form of overall spiritual matrix into which those who are exploring these areas fall, and are thus drawn into an altogether more complex game the dolphins themselves, or whoever might be guiding their actions, may be unravelling. If anyone reading this, who has also read Ted Mooney's book, would take a moment to consider the synchronies in their own lives leading to the reading of both writers, they may well find in these meaningful coincidences a similar experience of the telepathic matrix.

This opportunity for further explorations came sooner than we expected it. A colleague returned from the Bahamas with another odd and moving story about dolphins.

He had been staying on Paradise Island and one afternoon had set out to explore the rocky inlets with the hope of coming across some wild dolphins. He knew about our experiences in Florida and wanted to find out more for himself. After strolling through pine groves and past exquisite tidal pools, he suddenly came upon a truly incongruous sight. There, built into an outcropping of rock and forming a bay of its own, was a sweeping, curved, futuristic concrete quay, seemingly nothing to do with anything. Behind it rose the outlines of what appeared to be a large ramp again leading nowhere.

Somewhere in the back of his Twentieth Century mind wheels clicked and whirred. He had seen it all before. Struggling to hold back a mild shot of satori, his brain flicked through the options. A secret CIA loading station? A hallucinatory flash-back? A deja-vu? And finally, Dr. No — of course, they filmed the James Bond movie down here in the Bahamas.

Then, following the long curve and over a small bridge, also evidently part of the film set, there, right in front of him, a small sign reading DOLPHINS — THIS WAY, with an arrow.

Some minutes later, after a winding walk along a clear, blue lagoon, he and his lady stood watching the first two dolphins they had seen close up, making long and exotic love in the late afternoon sun. It was, by all accounts, a bewitching moment and one which rendered them both instant dolphophiles.

Consequently, when they returned to a snow-bound New York, it was with the news of four recently captured dolphins being kept in the private dolphinarium of the island's Britannia Beach Hotel. It felt to us like an ideal chance to take a period to study and observe dolphins in captivity and under conditions somewhat more accessible than those of the larger oceanariums.

Out of deference to the wild dolphins in Florida, we called our friends in Tampa to see if the fish were flying. No luck there, so it was green lights for a trip to Paradise.

CHAPTER FIVE

While circling to land in Fort Lauderdale the captain announced, in embarrassed tones, all routine flights to the Bahamas had been cancelled. A strike by government employees on the islands. Images flooded in of holing up in Fort Lauderdale and going on to Tampa to see our wild dolphin friends. But, after landing, we were lucky enough to stumble across a tiny airline using flying boats, as they were called in the heyday of air travel. Small 1940 seaplanes re-fitted to hold twenty passengers, and since **they** do not need airports, we could land safely in Paradise.

We waited serenely for our connection amid the chaos of the over-crowded terminal. One crackling message after another confirmed a growing hunch that much of the mechanical and electronic equipment in our sophisticated Twentieth Century systems was starting to fall to pieces. Everything shimmering just this side of metal fatigue.

Our seaplane, by contrast, managed the short haul with a juddering delight. The raw wind and air pressure cradled us as we skimmed low over the deep blue of the Caribbean. Invited up to the cockpit, I gazed down through the oil spattered glass at the Island of Bimini with its surrounding reefs and sandbars, thought by some to be the rising remains of Atlantis.

We found ourselves seated next to two of the architects for the Britannia Beach Hotel where we were to be staying. The one I talked to had a Master's in psychology and was very interested in our adventures with the dolphins, especially in the two UFO incidents. He remarked on the possibility of the images being projected directly into our brains, a concept we briefly entertained after the Manhattan sighting but had rejected because of the UFO's trace appearance on the photograph. He went on to remind me of the curious experiments conducted with Ted Serios: under control conditions Serios was able to produce images on film by concentrated thought. The question he raised by this startling feat is one that should warm the heart of any alchemist. Can thought itself produce effects on the highly sensitized silver coating of undeveloped film?

The intelligent interest of the architect and the fortuitous manner in which we were being transported had, once again, the echo of a higher hand at work. I fancy it was at this point we started to feel the presence of the dolphins again.

After an exhilarating hour the seaplane parted the waters of Paradise Bay in a spectacular landing, trundled up a small ramp and out we fell into a warm tropical evening. WELCOME TO PARADISE, the sign said!

The next day found us getting the lay of the land. And a very beautiful land it is too; intensely peacock-blue, transparent waters washing quietly onto clear-light beaches. A shallow canal drifts under two small bridges connecting the sea with the lagoon in which the dolphins live in their enclosure. Glades of pine and silver birch separate our hotel from the beaches and line the white, rocky banks of the canal. The hotel however was doing extensive building work and the ripped earth contrasted ominously with otherwise luxuriant surroundings.

The dolphin enclosure was a wired-off section of the larger lagoon, the long side of it formed by a timber platform projecting some ten feet over the pool. This platform also acted as a sun-deck to the hotel's swimming pool, about fifty feet from the dolphins. The dolphin compound itself divided into three separate enclosures; the one presumably used for the shows is surprisingly small and when we arrived held a single, large dolphin. The smaller of the other two wire-meshed pools contained three dolphins, two of medium size and one obviously smaller and younger.

As we drew near and caught our first glimpse over the wooden rail, all four of them became active, leaping in the air and speeding around their enclosures. We thought nothing of it at the time, just immersed in their beauty and, like our experience at the New York Aquarium, profoundly saddened by the woeful inadequacy of their quarters. Yet there was a real contradiction here; the dolphins' pool with its five feet-high wire fence was very evidently cramped. Watching the dolphins leaping, it became clear they could have easily jumped the fence and made for the adjoining ocean. What holds them here?

Could we be seeing the results of enforced activity? Might there be, as Rasta Bob, a Bahamian rastaman we were to meet later that day suggested, some form of subtle hypnotic hold established by the trainer which keeps the dolphins, in Bob's words, "spurred"? Then again, assuming them to possess the level of consciousness we found in the wild dolphins, might it also be possible the dolphins are fulfilling a function quite their own? A teaching role, diplomatic relations, a study of humans . . .

In support of this third possibility Rasta Bob had told us the dolphins in the enclosure remember individuals with extraordinary clarity. He said if someone were to feed them just once, or even watch their show a couple of times, and then return a year later, the dolphins would recall them and put on a special show or respond in some other appropriate way.

I returned to the pool at four o'clock the same afternoon to watch the feeding. A cursory affair by all appearances — I was not even sure I had seen it. Maybe I had not — I'd missed the feeding — but nobody else saw it either. After about fifteen minutes, the small crowd trickled off, grumbling in a puzzled way, leaving me and my companion, who by this time had joined me, looking down into the pool by ourselves. I remarked to her that the dolphins had started jumping once again when I had arrived. By this time I was starting to catch on; I'd had a chance to watch their activities over a longer period. They were mainly passive, staying in one spot for most of the day, merely breaking their routine for the two shows. However, as if to make my point, they had again started up their jumping and playing at the very moment my companion joined me.

Possibly coincidental? These matters are most delicate, but we had learned to go with the feelings and not block them with rationalizations. In some subtle way contact was being made. And, as we hung over the rail, the two dolphins in the larger pool swam slowly under us inspecting us with their sharp eyes.

There was an overpowering sense of contained excitement. They felt enormous, sleek, and utterly confident in what they were doing. They nuzzled each other, curling and bending their sensual bodies. One had ferreted up a large piece of seaweed and was prodding it into position, then swimming slowly alongside of it, allowing it gently to stroke his side. It was a languid and amused gesture.

They eyeballed us with a relaxed intensity, drawing

from my heart a warmth and a love, and an involuntary flood of "Oh, you're so beautiful . . . you beautiful, beautiful creatures . . . Oh, you beautiful creatures . . ." I recognized the flooding of emotions which often preludes a state of god-consciousness. Remembering this cooled me out somewhat, otherwise I think I would have been over that fence and into the water with them. But, easy does it . . . there is another way!

Later in the day we wended our way through the hotel gardens and, on my companion's suggestion, set out to explore a small group of shops outside the hotel grounds.

Rasta Bob was standing on the corner, touting his wares in the manner of Brothers the world over. There was a softness about him and a sensitivity with which he shushed mid-spiel his younger and more enthusiastic acolyte. No low-grade local was going to pass as 'sensi' with this tourist, that I knew he could see, and I appreciated his street smarts.

After shopping we re-joined Bob and the three of us sat on a quay by the lagoon as the sun set. He filled us in on the state of the Islands from street-level perceptions — often the most accurate — relaying a story common throughout the Caribbean, of political corruption in the face of largely passive public inertia.

"De natives here," he used the word with a disdain, as if separating himself from the Bahamian people, "Dey all after money now. Dey's imitatin' de white man, Mon. Dey wants objects an' tings an' dat's how de gummernt keeps dem down."

He said there was no major opposition party and an extremely small and, I would imagine, placidly revolutionary, socialist party.

"Dere's no chance dey get in power, Mon, 'cos de gummernt, it have all de money. Dey can tell anybody to leave, dey just give you thirty days to get out if dey don't like you." It was a familiar story.

The talk turned to dolphins. Bob, like many rastafarians, has a deep and vital appreciation of all animal life and, as is the case with most tropical islanders the world over, this extends to a particular love of dolphins.

"Where I used to live in Grand Bahama, Mon, de dolphins jump and leap in de bay right in front of me." He promised to show us the pictures he'd taken. "A lot of rastamen, dey is real close to dolphins. Dey incredible creatures, Mon, if you kind to dem, dey never forget your face. You know, Mon," his voice became soft and dreamy, "sometimes I go an' feed de ones over in de pool," he gestures at the enclosure some seventy-five yards further down the lagoon, "an' I fall right into dem, Mon."

I understood what he meant! I could also see he **knew** dolphins, in that quiet and sonorous way open-hearted people have of relating to them. His use of the metaphor "falling into" brought a reminder of my previous wonderings as to whether dolphins might project a telepathic web into which people of appropriate sensitivity are drawn. If this is so then it would likely be a function of the non-dominant hemisphere, the right-brain, with its accent on visual and non-verbal communication. And, from the vast quantities of ganja the rastas consume, it would be more than natural for them to be open to dolphin contact!

Bob went on to say the trainer had the captive dolphins "spurred"; hobbled and tamed. Not a pretty picture. Yes, he agreed, they could get out at any time but, "de mon, de trainer, he stupid, Mon. He keep de dolphins stupid too, by routines. Dey just follow de same routines day after day."

I pricked up my ears at this as it suggested a certain hypnotic form of brainwashing. By impressing particular and precise routines on open and receptive psyches, a person's soul can be deadened sufficiently to be made to comply with another's will. It is a method used with little awareness by almost all industrial societies, and in its more extreme forms, by prisons, armed forces and religious cults. Could it be the dolphins had themselves fallen under the hypnotic web of their trainer?

My companion slipped down to the pool a little after midnight to find the last of the night strollers had retired to the tempting crackle of the casino's one-arm bandits. Alone at the pool, all was quiet.

She tapped on the boardwalk with her feet and, within moments, the dolphins were leaping and blowing for her. She started to sing to them and they creaked and clattered back to her.

By contrast, my late night visit produced no antics at all. Two hotel employees, there by the time I arrived, scowled briefly at me as I scaled a gate with a KEEP OUT notice on it. It allowed me to be nearer the now-sleeping dolphins. They drifted slowly past me; scarcely a hint that they saw me. Then there was a mild commotion over on the boardwalk followed by the guard calling me back. I realized he was frightened something would happen to me and he would be held responsible. I wondered what harm he thought could ever come from dolphins.

Before I got up to leave, I looked down at a gently floating dolphin five feet below me and saw instead a large, steel-grey and very menacing form. Now where did **that** image come from, I pondered?

After climbing back out, I chatted to the guard and ascertained, as we had suspected, that the dolphins were fed only twice a day, at showtimes, Noon and 4 p.m. Odd hours to feed large mammals that, in all probability, eat whenever they are hungry under natural conditions!

This manner of regulating their food intake, apart from the routines, implies too, a certain level of malnutrition that might also serve to control their behaviour. Psychically open **people** are particularly vulnerable to poor nutrition, possibly we all are to a much higher degree than we give credit. Dolphins, functioning within a more subtle range of consciousness, may be extremely open to manipulation of this sort.

Every instinct tells me there is more going on here than meets the eye. Why did the dolphins jump and sing to my companion and not to me? Perhaps I went down there with some pre-conceptions as to what might happen, and pre-conceptions, as we learned in Florida, seem to promote a level of mental activity not conducive to the kind of consciousness needed for close contact with dolphins. Then again, it may have been purely because she was alone with them whereas I had guards and hotel employees present at the pool throughout.

The thought raised an issue which has pursued our involvement with dolphins from Florida to the captive ones in New York Aquarium, and now here. I inevitably come away from these moments with a sense, like an echo in the back of my mind, of some extraordinary game going on. As if it were **us** who were being led through our paces; the dolphins definitely knowing something we do not and they aren't letting on.

My companion arrived at the poolside next day for the Noon show. There were evidently no particular signs of recognition from the dolphins, who seemed pre-occupied by the feeding routines and procedures. I arrived half an hour later. By the time I hit poolside the two prize dolphins were being fed and doing some perfunctory tricks. Although singularly beautiful — dolphins cannot help but look magnificent wherever they are — they had a tired, dull-grey appearance about them. The show had nothing of the professionalism of a larger aquarium and the trainer exuded that curiously protective aura we had grown accustomed to seeing in dolphin handlers. The trainer in this case, one Duke, is a stocky, rather macho young man. From his evasiveness over the last two days, and he certainly knows we are here, I suspect he may well feel he is onto a good thing and does not want "any kinda writer down from New York" suggesting the dolphins might be immeasurably more intelligent than he!

Back to the show. The dolphins clicked and creaked at Duke while he threw first hoops, and then a large, plastic frisbee for them to retrieve. He rewarded them with small pieces of fish. Out came the beach balls. He threw one to a dolphin, who tucked it under his chin and swam with it for a couple of seconds before flicking it back to him. This happened two or three times and it became obvious that the dolphins have a very high level of control over the trajectories of the balls they throw.

Then one of the dolphins beaked the ball high over the wire fence into the smaller of the two remaining enclosures. It appeared accidental, or at least inadvertent on behalf of the dolphin, but I saw Duke was quite put out by this display of independence. **He** knew full well the dolphin did it on purpose. He grumpily tried to cajole the other two, as yet un-trained dolphins, to return the ball. He gestured impatiently to them, using first hand movements to direct them, then, as he realized they would not return the ball, he threw hoops, hoping they would get the point, only to lose **them** as well.

Something odd was happening and Duke was getting a little strained at the edges. It looked like he had reached the limits of his understanding and was getting aggravated.

After a while he gave up trying to retrieve the beach-ball and the show and the feeding careened on. We both clapped loudly whenever the dolphins did tricks, feeling at least a certain amount of appreciation needed to be communicated. The tricks melded into one another so it was impossible to say when one cycle stopped and another started. The small crowd, while obviously entranced by the dolphins, grew progressively more confused by the show's lack of parameters, and drifted off, lulled into their own hypnotic trances by a loud-speaker crackling out bingo numbers somewhere in the distance.

My companion and I moved closer together as the two dolphins swam under us, watching. We kissed briefly and the dolphins, on cue, started swimming and jumping together. By this time, the show had ended and Duke was cutting a cucumber for his own lunch. The two smaller dolphins were basking in their pool eyeing the cucumber speculatively. We retired to a couple of beachchairs about six feet back from the poolside railing. Remembering Florida, my companion suggested we pull our attention back from the dolphins for a few moments to see what happened.

As I lay back in the warm sun I found myself drawn into a deep meditation. I felt myself falling into the shapes of a dolphin, turning and twisting in the water. I relaxed into the feeling and realized, in an expanded field of consciousness, I could intertwine with that of a dolphin. There were moments, seemingly lasting for ever, in which I felt the water slipping past my skin, and then the thoughts started crowding in. I emerged from my trance with a distinct sense the dolphins had picked up on us and knew why we were there. Eager to record this brief but startling encounter I groped around for my pocket tape-recorder, soon finding I did not have it with me. I got up and put my shoes on in one motion and glided back into the hotel to fetch it. As I pushed open the door I had the clearest impression I was, in some way, still carrying the consciousness of the dolphin; he was inhabiting my body alongside me and looking out through my eyes as I moved ever-so-smoothly along the cool, blue corridors . . .

By the time I got to my room and started fumbling for the tape the impression faded.

I caught the four o'clock show, or not-show as was the case again. Duke just could not get those birds to fly! They would not even come out of the smaller enclosure when he opened the gate, so he threw a fishing net down into the water. A net! This is like hitting a bunch of children with a film of Buchenwald. I began to feel less compassionate towards Duke but I guess he has found the old net trick does it every time!

The dolphins disappeared at this point. Duke could not see them, and nor could anyone else. A girl standing on my right told me after about five minutes of nothing that she had seen a number of shows and this never happened before. I realized, if Rasta Bob's 'figgering' is correct and the dolphins have been here for as much as fifteen years, then they must be bored silly. Duke is holding them in a hoops-and-loops routine when they could be doing far more exciting things. Why not the whole lagoon instead of this mean little enclosure? Why not rides for the kids? Why not a real exchange between dolphins and all these people so profoundly affected by seeing them, but in a much more open and more nearly natural state?

Finally the dolphins breached and I managed to take three Polaroid photographs. I deliberately chose the

Sonar One-step wondering whether the dolphins would have any awareness of being sonared themselves. Duke was up in the helm, a high platform he uses for the big jumps, hanging out there like a piece of limp seaweed with a fish in either hand. The dolphins gazed up at him, creaking rusty doors and refusing to budge.

"Don't Sonar!" — loud and clear in my head. "Don't Sonar" again, insistent. O.K. O.K. I got the message . . . or was it my imagination? Go with it anyway and put down the camera. The dolphins immediately started jumping. Another coincidence? The show certainly went a lot better after I ceased taking photographs, and the dolphins swam around cuddling each other and grinning up at me.

By this time the crowd had thinned out and a young, robust American with a Southern twang edged up and told me how much he wanted to jump down into the water with them. I started to get interested — it sounded like me!

"My wife and I were out here at six o'clock this morning" he continued conspiratorially, "no-one around, just us and the dolphins. We could have slid right into the water with them . . ."

"Why didn't you?" I asked. He nodded and winked at me.

Before long Don had told me about some friends of his in Key West who had fed the same three wild dolphins every morning come wind or rain, and had done so for the last four years. It seems the dolphins were very shy for the first few weeks and kept their distance. Then, one morning, one of the dolphins beached itself, apparently in an effort to reach some food (a likely story!). Don's friends very carefully lifted the stricken animal back into the water and after that the dolphins totally overcame their reticence, even encouraging small children to ride on their backs.

While Don is talking, I get an image of dolphins assigned to make contact with as many humans as can tolerate a relationship with them.

By this time Don, who was down there on some company prize, was talking about Carlos Castaneda, saying he had read all his books three times. Good Lord! this banana must be some kind of heavyweight sorcerer — three times! I mean . . . Rasta Bob was heavy, but this guy . . .

We resolved to meet later in the evening. He ended off by saying a little peyote would certainly help in tuning into the dolphins. Mmmmmmmm . . .

After days of avoidance, Duke finally approached us, beguiled by a drawing my companion had done. He invited us into the enclosure and left us alone with the dolphins in the smaller of the pools. The young male and the more recently captured small female greeted us enthusiastically. We sat with them for almost two hours, talking gently and singing to them. They clicked and ratchetted back at us, their little black blowholes opening and puckering in the sweetest and most suggestive manner. They appeared to us somewhat like five-year-old children, innocent and utterly good. The very top of the mammalian evolutionary tree; trusting, helpful, mutually co-operative, yet, I had to admit, not possessing anything comprehensible or identifiable as self-conscious intelligence.

This was the closest and most intimate time we had been able to spend with captive dolphins and yet, try as we might, we found little real sense of parity. I wondered at the time if they were playing with us; I could **see** the intelligence, there behind the eyes, and I could feel a deep sense of inter-connection, but as for my hopes of a **significant** first encounter, nothing! My pre-conceptions of a truly intelligent and telepathic species disintegrated as I tried with my mind to project first this thought, then that action. Once in a while there was a degree of coincidence but it was never direct enough for me to grasp and hold onto. Nothing as pronounced as the interplay we had experienced at the New York Aquarium.

The clincher came when the male started playing with a beachball floating in the enclosure — possibly the same one Duke had lost earlier. The dolphin positioned it carefully and then tried to flick it up to me. I was able to see the focussed concentration that went into sliding the ball into the crook of his tail, curling back to look at it; then, with a twist of a fin . . . the ball fell a foot under my outstretched hands, bounced once on the platform and splashed out of reach into the largest of the compounds. Stalemate! The dolphin and I looked at each other for a moment like two schoolboys who had just broken a greenhouse window.

Ah! Ha! I pounced on the thought. He is going to signal the other dolphins to get the ball and throw it back. My heart leapt. What a test of reality! But alas, not a flicker. Yes, he did appear to creak momentarily but Brit and Tania, the two oldest dolphins, just kept on making love and did not take a blind bit of notice.

I would like to think, as I write after the event, that I cooly realized dolphins like their nooky as much as any of us and were not about to be distracted by two kids who want their ball back please, but no . . . my heart tumbled. Back to square one . . . just big, sweet, dumb, animals, I thought.

A slow surge of doubt must have started in me at that point although I was only briefly aware of the unhappy ripples on the surface of my consciousness before tucking the feeling away and moving on.

It was some minutes after that fleeting depression

and I was abstractedly watching the two mature dolphins making love. Their rhythms and the way they positioned themselves in the pool seemed to follow a discernible pattern; one result of this particular conjunction of circumstances was a series of mounting waves which broke against the stone banks of the lagoon. I saw, in a slow dawning of inspiration, how just such waves could shape coastlines and create sandbanks; how, in fact, much of their environment might be created simply as a **by-product** of their everyday life.

Come sunset of that day, I sat by myself on Dr. No's futuristic jetty playing my guitar gently into the swelling, evening wind. Idyllic as it sounds, all was not well. The depression which started earlier broke through again, flowering darkly in the thought of a totally Godless and mechanistic Universe; of life being exactly what it appears to be from the most superficial and materialistic viewpoint. A lackluster affair, grinding entropically into the coil of its own oblivion. No Gods, no angels, no aliens and no dolphins to light the magic of a wondrous future.

Someone walked briefly over my grave as I sat in the gathering darkness, now on our rainbow bridge, the small, weather-beaten, faux-vielle bridglet spanning the entrance to the lagoon. A great Chaldean head, carved from concrete by wind, water and natural dilapidation, frowned majestically down on me, still haplessly playing my guitar, now horribly mournful.

Then, as quickly as the mounting emotion broke over me, it was gone. On its way, presumably, to disturb some other poor soul before the night was out.

I wandered home, my spirit cleansed by this strange catharsis. Passing the pool, Tania, the female dolphin, was on her back gazing up at me. We locked eyes for a moment before she luxuriously and deliberately winked. I reeled happily back to the hotel room, picked up the book I was reading at the time, and found J.G. Bennett telling me:*

> "I am personally confident . . . there really is such work (certain work required for a great Cosmic purpose) and that there are people who understand it in a way that is not obviously visible on the surface. This means that there is in effect a **Twofold Life** on the earth. One is the visible, external life in which we all have to participate, and the other is an invisible life in which we can participate if we choose. In a sense one can say the first life is a causal life; that is to say, in that life, causes that exist in the past produce results that are being experienced in the present and which will be carried forward in the future . . . It is of course, called by such names as 'samsara' and 'the wheel of life,' and so on, but in a very simple way it is the ordinary life that we all live.
>
> The second, the other life, is non-causal, which means that it exists only in so far as it is created. It is the life of Creativity. Every creative act rightly performed is a means of participation in that life. And the search for creation is the search for that life.
>
> This is what is meant by the word **Work**, and when we talk about 'the work' or the 'Great Work,'

* *Gurdjieff — A Very Great Enigma* by J.G. Bennett

it refers to the invisible world which has to be perpetually created in order that it should be. And it is that we are called to if we are destined for accelerated completion. In order to enter that world, we have to earn the right to be in it, and for that we have to bring to it something **made by ourselves.** The first and simplest thing we can bring is our own capacity for work; our own capacity for transforming energy, and therefore for participating in the Creation."

And then, a few lines later:

"What is significant about the future of our world is the coming unification of every form of human experience. The Work is concerned with bringing people together and not with separating them. I am sure that this is a very visible characteristic of this Twentieth Century of ours. In one way, this obligation to unite stirs up most serious reactions, and therefore we have seen troublesome wars and hostilities and hatreds. But if you look behind all this, you can see they all come about because there is an urge to unite and not to isolate. One very obvious feature of this is the increase in tolerance that has come over the world, and the mutual acceptance by people, which is perhaps the most hopeful and admirable feature of our century, with all its depressing features."

Bennett then ends his excellent summary and distillation of Gurdjieff's most fundamental point with these stirring words:

"I am really in wonderment at the extraordinary power, the superhuman intelligence and consciousness that is directing the hidden affairs of mankind at the present time."

I knew, in the language of my heart, the answers to

many of my questions lay in these passages and the events leading up to my reading them. I envisioned a time when more people focussed their attention on the subtle balancing of these forces. I pondered on the Great Work of the sages of the past and on how so much of what we think of as 'reality' can be very directly traced to the actions, thoughts and perceptions of a very few men and women. Some of them indeed, like the shadow of St. Germain at the choosing of the American flag, having scarcely any objective, verifiable presence save 'being at the right place at the right time.'

Is understanding this a key, in some form, to comprehending the role of the dolphins in the creation of our shared environment? Might the dolphins be playing an essential part in the creation and upkeep of this invisible world about which Bennett speaks? Some spiritual traditions, the Urantia cosmology for one, talk of highly spiritual entities living so close to us they consider it unfair and prejudicial to their safety even to reveal their identities.

In the light of what I was discovering, could it be the dolphins to which these traditions are pointing?

We met Don, the urban sorcerer, together with his wife at about 11 p.m. on the night of the Vernal Equinox and rapidly entered the closely-shared consciousness which often develops in travels or chance meetings when, knowing we shall probably never meet again, we drop our normal barriers.

Having met at the dolphin pool, which may well have contributed in some mysterious way to the rapidity of the psychic upstepping we all experienced, we walked slowly down to the small bridge at the mouth of the lagoon. The white sand path snaked alongside moonlit water, the buzz of hotel life diminishing with each step nearer the sea. My companion had announced the day before, quite spontaneously and in a voice carrying the unmistakable authority of a channelled message, that we should be on the little bridge at midnight.

The bridge still held memories for me of my curious and sudden despondency. Perhaps this may have added to the feeling of intensity now permeating the place. The massive Chaldean head thrust high into the moonlight and the dilapidated paving crumbled under our feet. Over all hung an air of impending strangeness to come.

I could see Don and his by-now clinging wife were getting progressively more freaked out by the emanations of weird portent. Before we knew what was happening they were both off and scuttling back up the path through the trees to the apparent security of the casino. We were left alone as midnight approached.

I found myself possessed by a mountingly powerful force, words pouring like a current from my mouth, alternately sitting and jumping up, pacing backwards and forwards, delivering an impassioned discourse to my companion and who knows what other, more invisible, entities.

Then, silence.

And in those few serene moments we both heard, with the ears of the Spirit, three distinctly separate waves, each with its individual and wholly defined characteristics, roaring up the canal, under the bridge upon which we stood, and on, into the lagoon in which the dolphins were held.

It was precisely midnight. The words teemed out of me again.

"The circuits are opening! The circuits are opening!" I sung and leapt and laughed, barely knowing what I was saying. Images played in front of my eyes. I saw the precision with which the set-designers of the movie Dr. No must have planned and engineered the jetty and escarpment to catch these three waves, on this particular evening, albeit years after the completion of the filming. No matter! I knew intuitively, from other, acausal levels of reality, that the hands of those engineers had been guided by some form of angelic coordination; it had all been planned in the Way of the Spirit to capture this one moment and relay it on to the waiting dolphins. It was **us** who were the eavesdroppers!

I knew with metanoiac certainty that the dolphins, and the entities I was now starting to feel were guiding them, are totally involved with the sustenance of the more subtle energies at work on the planet and, as such, the waves conducted information essential to them.

And more profoundly still, I knew in my heart the isolation within which our sector of the Universe has for so long languished, is even now drawing to an end. The circuits connecting all the inhabited worlds of a teeming Universe are, for us and our small System, by some act of Divine mercy, in process of being re-established.

Confirmation of this exhilerating possibility did not come until we encountered a more direct line of contact with Angelic personalities some months later. At this point it remained pure revelation, facilitated in all likelihood by the geomantic and geo-magnetic peculiarities of this small island set on the edge of the Sargasso Sea.

For all my watching and studying the dolphins in the hotel enclosure, my first moment of intense intimacy occurred not there, but at the Nassau Seafloor Aquarium. It was a blooding into the next stages of Trust in the Inner Journey. An initiation.

We had set out for the aquarium on an overcast day which generally added to the air of depression hanging over the poorer areas of a hypertrophized economic boom gone bust. The quietness and lack of tourists suited us just fine however, after the razzle-dazzle of hotel life.

The dolphin pool is set high in a glass-sided oval tank and surrounded by a well-organized garden of tropical shrubs and exotic flowering bushes. The scents hung thick and the conch shells lining the paths shone with the recent rainfall.

In the tank were two female dolphins. A mother and daughter. They were caught together when the daughter was a mere few days old and the shock of the capture had dried out the mother's milk. It had been touch and go for six months but finally everybody survived and now the daughter is the more advanced of the two, learning faster and more effectively than any dolphin previously held at the aquarium.

Indeed, the two performed exquisitely together for the small crowd. Like the hotel dolphins, they seemed fixated on reward systems and the same kind of routines we had noticed all week, but their skins appeared to be much clearer and healthier. We wondered whether this was due to the form and level of sexuality at the two places. The presence of the two females without males went a long way towards lifting the emotional pall hanging like twilight over the dolphins back at the hotel.

After the show had finished we stayed near to the pool waiting for the trainer to disengage. I was casually dribbling my hands in the water and watching the sea-lion resisting all attempts to be moved out of the pool area.

Something stirred in the water near my hand. It was the young female. A sleek snout, and one very bright eye, hanging back just a few feet beyond my hand.

When she was assured of my attention, she opened a large mouth lined with rows of diamond-pointed teeth. Then, with extraordinary intensity, she moved deliberately towards my fingers. A moment of fear washed through me, racial heritage of a predatory species, before she soundlessly closed the gap and rested her chin in my hand.

It felt like a profoundly conscious gesture on her part. She knew exactly what she was doing and how I was likely to react to her. Perhaps she'd even seen JAWS at the local fleapit. But she'd pushed ahead anyway in an action resonant with reciprocal trust, as any are bound to be in these early meetings of essentially alien races.

My last encounter with the dolphins in Nassau also carried with it something of this trust and intimacy. It happened, not in the aquarium, but in the cool glow of a low early-morning sun at one of the hotel's inner enclosures. It was also, however, with another young female dolphin.

The hotel was still asleep; no one around and no sound but the slapping of the waves and the calls of the early birds catching whatever they catch in these climes. I sat, my back against a timber post, legs hanging over the edge of the small, wooden platform, playing my guitar to the lone dolphin a few feet below me.

She seemed fascinated by this odd break in her routine and certainly **appeared** to listen, pushing her head out of the water and watching me with half-closed eyes. Knowing dolphins have no ears as we would recognize them, I imagined she must have been opening herself to receive the sonic vibrations of the guitar strings. She made no attempt to sing back to me however, as I hoped she might.

When I had finished a twenty-minute stretch of

music, the echo of the harmonies hovering in the air between us, she waited for a short pause and then dived sleekly to the bottom of the pool, re-emerging a few seconds later with a small rock between her teeth which she reached up to give to me.

She repeated this three times in such an obvious gesture of gratitude that I knew something of extraordinary intimacy had passed between us.

CHAPTER SIX

My deep personal desire for close physical contact with a dolphin, mounting to almost an obsession since I first met the wild dolphins in Florida, seemed to have been fulfilled in these two magical moments. Both incidents, one with the music and the rock and the other in which the young dolphin cupped her chin in my hand, were suffused with tenderness and reciprocated emotions.

It was with these encounters still resonating in my soul that we sat in the gardens of the Seafloor Aquarium to ponder on our experiences so far and exactly where they might be leading.

My focus of interest from the start had been on the nature of telepathic communication and intelligent interspecies contact. Indeed, it was primarily in the light of these two factors I had kept such careful notes — I did not wish to miss one nuance. At that point I had only a very limited understanding of what 'interspecies contact' actually meant, and more significantly, what it would come to mean. The extra-terrestrials had shown themselves briefly and there had been intimations of the angelic reality with its vast array of spiritual and semi-spiritual intelligences. If I'd have been more sensitive I might have also perceived, as I do now, the workings of still other intelligent Beings, the Devas of the mineral and vegetal realms, known too as the Nature Spirits.

The integrity of these adventures lay in what we saw as following the 'dolphin trail.' It was a question of hanging loose and pursuing the signs as they emerged. Thus, it was in Nassau that we received the next stage in our continuing preparation for the revelations which were to lie ahead. We found ourselves plunged into encounters with an assortment of people who, by nature of their backgrounds and cultures, were sufficiently divergent from ours as to make the rapport seem, for all the world, like modified and practical forms of yet other aspects of interspecies contact!

I knew there were ancient lines of knowledge, mostly occult, and passed on through oral traditions and carefully-tended bloodlines, because I could see it periodically surfacing in societies and individuals throughout the recorded history of the planet. Yet I had no real inkling of the Network of Light, what it might be and how it functioned. The dolphins keyed us into it, that much was clear, since it was becoming more evident to us that they too, are instruments of higher powers. Agencies becoming progressively more visible to us as the adventure was unfolding.

We had decided on impulse to walk back from the Seafloor Aquarium, gasping before we had travelled two hundred yards in the monoxide haze.

We stopped on a small side-street and were photographing the surprising sight of two quite different trees growing, but one **inside** the other, when from behind us came the by-now familiar "Hey Mon." We turned to see a striking six and a half foot rastafarian with full dreads curling and coiling medusaic around the head of the black goat-God Pan himself.

We were ushered into a diminutive Bahamian

dwelling — I want to call it a shack but the Bahamian people have an unusual relationship to the physical and 'shack' would be a thoroughly consumerized evaluation — largely bare rooms with skar music reverberating every bit as much as if the house itself were the speakers.

Leaving my companion in the front porch to talk to a goat, our new host, Mark, beaming dark delight, led me through the house and into the back yard. We sat down on concrete blocks between newly planted beans and peppers. Mark does not eat meat. He tells me: "We no business killin' de animal, Mon." I agreed as we both rolled up our various herbs. My New York jay looked ridiculously ineffectual next to a spliff the size of a forefinger. Still, he was noticeably uplifted by a couple of blasts of Panama Red while I tried to huff and puff up a storm like I'd seen the rastas do in Jamaica. We both coughed and spluttered, laughing with Thomas Cheech at the goodness of the herb and the way it always makes us cough.

The barriers dissolved and I saw Mark carried the Spirit in a deep and resonant way. He knows the Spring of the Heart has come and that these are strange and miraculous times . . .

My companion joined us as Mark was explaining his own interpretation of the Jupiter Effect, the multi-planet conjunction that has been gathering momentum these last couple of years. He told us how this is initiating new forms of spiritual contact.

"When de planets in dis conjunction, it make it possible for de Spirit to come straight down. Dat forms a triangle, Mon, between you an' me an' . . ." he pointed up and out and I could hear the Spirit start to talk through him. There is a special quality about words spoken in truth; words from the heart reverberate with their own meaning.

He laid out some of the rastafarian cosmology for us, and I heard too the deep inner meanings. He was teaching, not evangelizing.

Solomon and Sheba, so it seems, bore progeny and it is from the first-born of their children that Haille Selasie was the direct descendent — this was in reply to my question as to the role played by the Ethiopian emperor in rasta thinking. As Mark spoke I could see the spiritual instincts of the rastafarian train of thought. In tracing the House of David through to Solomon and Sheba, they were definitely tapping off a rich union of the two most powerful lines of knowledge available on the planet at the time. The Queen of Sheba was herself Ethiopian and her natural knowledge was well-matched with the high mysteries of the Egyptian inner circles. It was the product of the genetic memory of a pure and powerful blood-line.

In this manner Haille Selasie, the Lion of Judah, King of Kings, Emperor of Ethiopia, and the direct blood descendent of the House of David and Jesse, is believed by the rastafarians to be the bearer of the second coming of Jesus Christ.

I have to admit this bit has always bothered me about the rasta belief system. The Emperor, for all his titles, never felt to me to be quite the part, and his recent death came and went without undue spectacle. Yet, within the subtle reality of the Great Work, there is a substantial point being made. Blood lines **are** important. Natural knowledge is largely carried by genetic endowment, a factor modern science has yet fully to appreciate. Possibly as a spiritual reminder of this, we have the rastas, some of the most righteous, honest and downright fine people you can hope to find, all united around this one rather self-effacing man.

The second coming of Jesus Christ, I think not; but

the Spirit of Truth, a teaching Spirit sent, as I believe, to reside in our hearts, plus the preservation of an invaluable line of knowledge, both coming together at this point in time, reflects a form with much continuing vitality.

This feeling was increasingly borne out as the Spirit continued to talk through Mark. He spoke little of himself but I gathered he is in his late thirties, the first-born of seven to a reasonably well-to-do Bahamian businessman. He lives with his sister, who essentially keeps an eye on him, in a house within eyeshot of where he grew up. He loves his mother and appears, in turn, to be loved and respected by all. He is the local saint, sage, shrink and comfort to the needy. A scholar and a teacher. Wonderfully and woefully impossible on the physical level, I suspect — there are rusted pieces of engine strewn down beside the house — but wise men often have a reputation for having trouble with their shoelaces. We cannot all be good at everything.

I inquired about the first-born. Why had he placed such emphasis on it?

"De first-born belong to the Almighty . . ." loaded words, as he rumbled on about door-posts and angels of death, and I saw my companion nodding in the corner of my eye. I realized we are both first-born too.

The 'chance' meeting slipped into a new gear. The sky darkened and thunder crashed and crackled around us as we made for the "shop," a small, bare, room adjoining the main house. A fire was started in a corner of the concrete floor and soon soft smoke wafts under a half-open, top-hinged shutter. Some magnificent Turkish coffee is produced, "straight off de ship dis mornin' Mon. A gift from a friend." Mark is evidently well-placed!

Rain poured down outside. Having lived in Nassau

myself, some years ago, I had seen my share of tropical cloudbursts — add in a couple of monsoons and a hurricane in other lands — but I have never witnessed anything like the size of those raindrops. They were so improbably large there appeared no reason why they did not disintegrate into a thousand normal size drops. They were cutting white swathes through the air like a cartoonist's pen. We danced and sang and gave thanks for the best and richest Spring we had ever seen.

The rain stopped as startlingly as it had commenced so we went back outside to sit under the shining, wet trees. A quiet joy settled over the three of us. We basked in the mutual understanding and contact; little needed to be said as the Spirit fell across us and the gathering shifted perceptibly into top gear.

With language falling away, Mark and I looked deep into each other's lives. Within this silence I **knew** suddenly and in a moment of very clear inspiration, that the Baraka, the spiritual power, of Haille Selasie, upon his death had passed over to Mark. The recognition was wholly intuitive but undeniable to the eyes of my Spirit.

Mark knew this. It had been recognized before, he told us, and related his own lineage through the tribal complications of Jacob's many sons. We stood up and shook hands solemnly at the truth of these remarkable revelations.

We talked, laughed and enjoyed each other long into the tropical afternoon, Mark a mine of natural wisdom and esoteric traditions, a true prophet for his people.

Dusk was falling when, with open hearts and a common agreement to see our mutual Father's work progressed, we set off again down smugglers' streets, back into town.

By the time we stepped onto Bay Street, Nassau's main drag, Eli, a pleasantly restrained street hustler we had encountered earlier, bounced up to us again. He later confessed he had kept his eyes open all day for our return — his heart had told him ours was, in some way, a significant meeting.

I rapidly checked him out as we walked together to a restaurant. Yes, it was a fine Spring he felt, **and** good for the flowers. Well, that is how much **he** knew. We sat down in a local Greek eating house and within minutes the emotions were starting to flow. The guy was crying behind his hand, sheltering his face from the rest of the diners. I could see he had been waiting ages to get this lot out. He has been all over the States, a roadie with Thin Lizzy and some other name groups and regarded himself as well on the way to becoming an independent producer. I believed it. He had a lot more class than your average street hustler. Yet he gave it all up to return home; he did not know why but he just felt like the prodigal son. Once back, he simply could not leave.

Apparently he hails from one of the most powerful families on the Islands. Fifteen sons. "An' I the fifth-born, Mon . . ." — I could play at this game.

"Who is the first-born?" I asked him. A new welling of emotion behind the cool.

"Henry, he de first-born. He de biggest wheeler-dealer on de Islands. I love him, Mon, he de finest in de world. He absolutely straight . . .", more tears and then some insight into the ways of the people hereabouts. In what I shall probably remember as the 'no. 5 son rap' Eli went on to describe how his eldest brother had been testing him in some of the harshest ways imaginable but, Eli was quick to point out, always with love, to temper his metal and weld a tight and powerful family unit.

Henry apparently "can't stand the sight" of poor Eli at this point, although the latter claimed he has remained absolutely loyal throughout it all. The agonies and ecstasies that get played out on these islands!

We offered to accompany Eli to see his brother, who sounded like a man with whom it was well worth spending the time of the day. Plus, I'm told, he plays a fine guitar. Eli, predictably, is terrified by the whole prospect — we are the only people who have ever called his bluff. He oscillates wildly between the joy of seeing his brother again and his fear something will go wrong. He doubted whether Henry would ever play the guitar for us, announcing with prophetic certainty, if he did it would be a sure sign all would be well. After one last tearful deliberation he decided to go ahead and risk it.

The cab hurtled back 'over the hill' which, in island parlance, can either mean the Bahamian section of town, or the ghetto, depending from which side of the hill you are looking at it. It was quite dark by the time we climbed out in front of a small dilapidated house. Splendidly robust tomatoes stood with some other fruit on a wooden shelf in front.

I approached the front door. I felt Eli holding back, half-hoping his brother was not there. A curtain shivered and a muffled, hoarse voice called through the thin wall: "Use de back door — use de back door." We ducked into a low shed open to the yard, another door at the far end leaked a pale green light. As I walked into the small, stuffy room I sensed I could have been in any one of those million small rooms where groups of men huddle and talk and bargain endlessly in the sparkling, reflected light of the whitest lady of them all. Nine men and women were frozen still, caught for a moment in their own private paranoias at the sight of my white face in the green underwater light of a 40 watt bulb. I have had warmer greetings!

I intuited my business was with the man in the center with the face and bearing of an African king. He wore Judo gear and a Ché beret.

"Your brother tells me," I waved somewhere behind me, "We have some interesting things to talk about . . ." I opened. Pause. A flicker went through the room. His voice a little higher than I expected.

"As you can see, I busy Mon, maybe tomorrow . . ." he countered, searching for my intentions.

"We've come a long way — we can wait outside . . ." I half-turned as I felt the host rise in him. He backed down gracefully in the space I gave him and led the way into an inner sanctum. His own room, I assumed; I felt the specialness. Eli, finally made it into the house, and the others, dissolved into a next-door mumble.

We settled almost immediately into the language of the heart. The talk veered through island politics, the psychological state of the Bahamian people, the difficulties of a new country, wild dolphins, and of deals made and un-made. Now, at thirty-eight, he had found himself at the top of his chosen profession. He spoke of his conflicts about being 'in de business' and yet having a deep-rooted feeling of wanting to do something for his country. His leadership potential seemed unassailable. His obviously singular poise and deep intelligence, gained from experience, seem to have blended with an utter integrity. As Dylan put it, to be an outlaw, you got to be honest.

Not surprisingly, he knew Mark well and placed him as one of the few men with whom he can share his heart. His perception and analysis of Mark and all his sweet and saintly ways, was probingly accurate. Mark, he maintained, needed to learn everybody cannot be trusted and that he currently had some people around him who were pulling on his power. Any help Henry could have given him was promptly dissipated by these hangers-on. And, most importantly, Mark had to see this for himself.

Music came up and we both grabbed simultaneously for guitars and dialogued happily, twisting and turning around each-other, at first tentative, then stronger. I played free-style and I noticed he let me take the lead — an elegant courtesy by a true king. Playing against free-style can only happen right when there is real magic in the hearts of the musicians. It was one of those moments jazzmen wait for, early in the morning when the last drunk has been thrown out, and the very best herb is produced. As Henry said, "You can touch de notes wid you feelin's. Lotta people dey can play good wid deah fingers but you got to be able to touch de music itself." And he was **good**! Sweet and melodic under all the contained power and sternness bred of ghetto martial arts and fourteen brothers.

Elizabeth, his lady, came in, young and relaxed — little sign of chauvinism in this relationship. She was a young sorceress learning her craft. He knew it and she knew it, confirming it later by telling us how much her power had opened up since she had been with Henry. She lay back comfortably on the small bed while Henry left the room to fix me an orange juice. My companion slid over to her and they spoke quietly together. I realized they were talking about how to mend Henry's poor lungs. He had apparently shown some weakness in his chest in spite of a year and a half of abstinence. I knew I was witnessing a passing of occult ways between magical women, and that this event would start a morphogenetic wavefront of such passings, to the benefit of all. The wisdom of Sheba re-emerges when necessity and opportunity meet and balance.

The small room, with its carefully chosen ornaments, swam again with the gentle winnowings of our guitars — the revolution, here on this tropical island, was to be one of music. "The people die from lack of vision," says my companion, and Henry heard it good. A Pied Piper from better days, he will lead his people, in music and song, back to the Wisdom of the Heart.

We played and sang through two hours of intimacy of the spirit. Perhaps they will say in the future how those hours were days and the knowledge passed, the wisdom of the ages. They will doubtless weave their songs of the night when the first-born sat together and planned their plans — the day, the moment, that revolution, or more hopefully, evolution, was reborn on the island.

And then, centuries later, one more toke on one last spliff and a taxi was there to undulate us back into town.

Eli was cool and a little sour in the front seat. Conflicted by his joy in seeing his brother again, he felt bad, I guess, because his voice got the better of him earlier, back at Henry's house. But he was just plying his art, learning his craft . . . hit me for a Twenty!

. . . I gave him fifteen. After all, I'm Scottish . . .

I told him not to hustle his friends as we all piled out at Paradise Island bridge, and he beamed proudly back at me. I loved him — sure has some guts! If the positions had been reversed, I am not at all sure I would have got it up to do the same!

We walked on high over the late-night bridge, humped up over battered boats and mounds, literally mountains, of abandoned conch shells. The best and the worst, all mixed together in Paradise. The lilt of the place. The Spirit bubbling and eddying just under the surface of island life, thrust up through one man's soul here, a woman's heart there. The constancy of the sea; its ever-present whisper, white noise on anybody's track. Then, the pure foolishness of the frenzied imitation of pale-face ways. Cars, for instance, endlessly long, sixties and seventies gas-guzzlers spewed out of a mindless Detroit in better but stupider days, roaring to and fro, to and fro; low-grade monoxide fumes so thick you walk into a wall of dis-spirit, head-aching, numbed senses; bougainvillea, now grey and tarnished with fine, muted powder. The casino, with its crazy, confused standards, drawing the ever-hopeful into a web of pettiness and false values.

But soon, all this will change. The worst is over and tourism is dropping off radically in the face of global inflation and local greed run wild. Fertile turf for an evolution of the Spirit, and one that will happen quite naturally within the rhythms of the people.

Mark was sleeping when we arrived the next day. In the backyard stood, or rather, swayed, a new-born goat. AIO, sweet, protective mother, dark placenta still hanging off her, mehed plaintive-pleased by her living tribute to this new Spring. The five dogs, intensely curious although not altogether disinterested in the afterbirth, watched her. Young 'Half-AIO' snuck deep under the sagging, jilted floor of the wooden hut that is their night home; AIO unflinchingly butting any dog sniffing too close. Lizards snit and scrithered, tail-flicking, head-swaying, still suddenly for a reptile age, watching alert in minutest detail. One puffed up sacked throat on seeing me — another who dares to terrify?

Mark emerged to a new goat. Goat-God himself, laughed and purred baso-contenticus. Then his sister squealed, giggled, confident in matters of birth and hopeful motherhood, cleaned up and cleared away, lifting and propping the recently fallen corner of the goat-hutch with a concrete block.

My companion drawing the flowers and the trees — pleasant, contented mood of her own. We puzzled out just exactly how palm trees grow, each leaf creating its own net of coco-fiber that supports it throughout its life.

The mesh starts out fine and soft, delicate as woven cloth, then, as it ages and becomes nest to ants and lizards, it gets coarser and unbelievably strong.

Mark has made earlier reference to the Soursop Tree, the leaf of which is excellent for frayed nerves. We had plumped for coffee however, at the point when it looked like he was preparing to make tea from the leaves he was plucking.

All was righteous in the garden. I watched two lizards skitter face downwards, eyes glittering, looking at me first with one side, then another. I moved to where my companion was sitting, caressed her back and gazed appreciatively at her informal study of local flora. Hunching behind her and finding myself turned on by her scent, the morning, the garden . . . I nipped her shoulder in play but bit deeper than I had expected. She pulled back, gasping in pain and surprise, her soul-body drawn shudderingly back into the physical. Frayed New York nerves sent echoes of 'the bite . . . the bite . . .' reverberating through our shared neural circuits. Out of the corner of my eye I saw the others stiffened against the psychic storms. I held on for dear life, buffeted by ancient torrents until the warm Spring morning eased the howling of the hounds. Mark, ever-sensitive to the movements of the heart, served us Soursop tea anyway and we all laughed and surfaced together.

We made moves to visit Brother David, a young rasta with whom Mark feels very close and open-hearted.

Wending our way once again through streets lined with broken buildings and wrecked cars, the worst clutter of consumerism, we passed the general store owned by Mark's father. As we stroll by the shop I have a hunch the tall, dignified black man sitting on the corner of a '69 Pontiac is his father, although Mark, somewhat mysteriously, does not confirm it until we are all some paces past. There had been no perceptible acknowledgment from him to a brief nod from the man.

I asked him if his father was unhappy or disappointed about the nature of his first-born's chosen life and whether he ever expressed a desire for him to take over the store. Mark laughs, replying "No Mon, dere is plenty of other brothers" . . . and I pondered briefly on how it must be to father a holyman. The stresses and strains common to almost all cultures, especially to those subject to European colonial values, must have reared their beefy little heads; all expectations and hopeful dreams invested in purely material and physical betterment; the thoughts of a child's education the likes of which the parent never had; belated and unsuccessful attempts to affirm a consumer rationale against more basic levels of good judgment.

The lives of the older of the generations alive at this point in history, those between sixty and eighty-five, have surely fallen across some of the most powerful global upheavals the world has ever seen. Having participated in two world wars, their aims and ambitions, of necessity, have had to involve an over-attention to the material planes. The building up of resources only to see them tumble in recessions, depressions, and still more wars, has created a shocking sense of dispirit among those not yet aware of the extraordinary changes upon us. Amid this thought, I saw the quietly good way in which Mark's father has resolved this issue for himself: there was a maturity and a balance in the silence of their greeting.

Mark spoke to us about the problems of a prophet in his own land. Never an easy assignment! He told us a man must be "uplifted" by prophets from other places.

"Someone brought Bob Marley to see me — met

him before de show." They had liked each other immediately and there must have been a spectacular passing of the powers, "an' when he stand up in front of de people, he point directly at me an' he say, "Dat brother he **know**!"

Mark called it an annointing and the thought triggered in me another spontaneous vision. I saw how, by addressing the divine fragment in others, we ourselves become touched by God's Grace. Thus we are starting to catalyze, in these strange meetings, the group-souls of the various cultures we are drawn to visit. Once again, I intuitively felt Mark knew this, or was perhaps even generating these revelatory flashes in some manner.

We walked through suddenly-holy streets and after a few minutes turned into the mechanics yard of a Mr. Fussell, pronounced Foosell. After some window banging from us, he stumbled out with a hangover and the story of his lover twelve weeks pregnant. Worse still for Mr. Fussell, she is married to a man capable of murderous rage. Fussell's ten year secret affair with her has already yielded one child which, by the glimmer in his eye, he clearly felt is his. The husband, already unhappily suspicious but owning the child anyway, has threatened to kill the mother if she leaves.

I asked Fussell if they loved each other.

"Yes Mon, an' she ready to go. She say she want to leave him before dis happen." I watched while his life swam in front of his eyes on a thin and attenuated hose. In the silence I became aware Mark was bringing me here to see the chronic, day-by-day palaver of small island life.

Fussell started straightening up, aware in a surge of certainty, the moment had come. It was time to front the situation up and claim his manhood. I had no doubt, as we all touched each other's hearts, Mark himself had been encouraging Fussell to go do it and was using our presence to confirm the wisdom of following the path of love.

The interlude was brief and self-closing. We walked slowly out of the yard under a noonday sun, out into the street and the heady hues of recently damp vegetation. Soon Mark pointed out David's truck in the distance, up a low incline, and the sight encouraged us over the final stretch.

Prophet David stood in the low branches of a spreading tree, with ten or twelve other young Bahamian rastas, caught still for a moment by our unfamiliarity; draped, crouched, and squatting on the ground and in among the other branches.

We were introduced, with much hand-shaking, full eye contact and rastafaree exchanges. I did not immediately identify David; he turned out to be younger than I expected and his powers were still hidden. Within moments Mark fluidly separated him from the others and we walked over to a shack, part of the small compound on the far-side of the road.

Mark explained that David is a drummer in the Haitian tradition emphasizing his Bahamian mix of spiritual heartfulness tempers the misuses and excesses sometimes seen in Haitian voodoo. Indeed, somewhat in confirmation of David's redemptive ways with the Old Powers, my companion had pointed out a Haitian church nearby, boarded up with planks across the door nailed in an unequivocal blast of good riddance.

David showed us into a carefully tended little room set off from the main cottage and we sat, the four of us, crosslegged on the floor. He played 'A tribute to John Lennon,' a rasta scar album, on a sparkling new collage of stereo components. After a few moments at the standard deafening volume, our sensibilities rattling around the corners of the room along with the sound, David turned it first down, then off altogether. Some internal wrestling seemed to be going on, Brother David being a prominent local deejay, but his interest alerted when we started talking about loud, canned music as behaviour modification. There is a deep love of live music on the islands and the keener among the younger ones are getting on to some of the abuses born of rampant consumerism.

Before long the pipe of peace was passed around and soon the room breathed with a new resonance. Mark explained:

"Prophet David is a Dreamer. He very intelligent an' he think long time on somethin', then he dream a dream an' de thing happen."

David was lying back, the dust settling in his imagination. I could see the forceful and questing natural intelligence of the shaman breaking through. I found myself witness to something truly extraordinary; a young holyman, acknowledged by his own people (rare enough!) as a prophet, who can dream down reality into being.

I have heard of such special people in certain native cultures and intuitively knew it is true and can be done. I whelmed on the concept: reality, consensus reality, the one we all think is 'going on' out there, can well be viewed as a conglomerative construct. The product of all our individual realities. Within every cultural group, or more literally, group-mind, there are influential individuals whose reality is in some way so coherent it sets its stamp on, and therefore, shapes consensus reality. This is clearly evident in the grosser realms of the material world but where its impact is most strongly felt is within the more subtle levels of dreams and the imagination.

Sometimes these people are known and identified; Francis Bacon, Buddha, Christ/Michael, Confucius,

Albert Einstein, Leonardo, but most often I suspect, the true changers are not seen. They resonate with a higher truth, gradually pervading those around them in ever-widening circles of morphogenetic wavefronts. This effect is seldom consciously perceived by them or those near them and perhaps we all add our bit to the nature of agreed-upon reality in our own measure. Nevertheless, these people become fulcral personalities in the subtle realms and it is their hearts' wishes which largely co-form the nature of reality. There is little doubt it is these people to whom our young, alien mouthpiece was referring when he talked about the hidden vote-carriers.

Only in cultures where there is sufficient silence and faith in the miraculous workings of the Universe can these men and women be seen. They are the prophets and fools, the saints and the seers — those who ponder the deeper issues in their hearts. The poets too, don't forget the artists and the poets!

And this dreaming prophet! He will surely turn out to be a prime key in the betterments to come on this Caribbean holy land.

I broke the spell of our shared thoughts, realizing in the moment of changing to a verbal mode, this vision of interlocking concepts was somehow produced **by** our mutual interaction and known and understood by each of us.

I asked David how he would like to see the future of the islands unfold. My question brought the unsaid more clearly into focus and I felt it would help his dreaming to articulate his thoughts.

Mark nodded in the corner of my eye. The prophet absorbed the question, percolating it in his heart for a long pause. I added that given his role of dreamer, this was a deep question and could be answered in a few days should he so wish. He thanked me with his eyes but motioned me to indicate he would take a preliminary swing at it. His reply, richly and quite unrecordably woven in rastaman/voodoo verbal brocade, described with astonishing lack of judgmental attitudes the soulless stalemate of the rich and the poor on the islands.

His spiritual wisdom focussed on the main issue which has been surfacing with a real virulence since the islands gained their independence in the late sixties. Under the reasonably benevolent British-raj colonialism, the status quo of rich and poor remained relatively stable. The rich were white and the poor, black. There were a few better-off Bahamians but they had nowhere near the wealth of the Europeans, and later, the Americans. Consequently the problems of money and materialism never fully presented themselves to the native Bahamian.

Now however, with fifteen years of self-government behind them, there are some very rich Bahamians and, corrupted as some of them are bound to be by sudden immersion into 'the good life,' they gain and exhibit this wealth at the expense of the other islanders.

David, surprisingly enough, did not call for a redistribution of wealth but for more understanding by both sides of the preferences of the other. For a young and aggressive male of his revolutionary potential, he asked solely that the rich should not seek to impose their materialistic values on those to whom the spiritual ways are more important. He saw no problem with them being rich, but not at his and others' expense. He asked merely for a mutual respect and an appeal to both parties to meet with their heads held high. He ended with a well-shaped call for sanity on the imposed legislation on their beloved herb.

"How can dey make a plant illegal, Mon? It's Jah's healing herb an' no mun have de right to tell me whether I can smoke it or not."

Mark nodded enthusiastically at the intelligence of the reply, and indeed, David's consideration and subsequent answer showed a wisdom and breadth of vision well beyond his seventeen or eighteen years. It contained heart and insight and was undoubtedly barely the tip of the iceberg he would soon be dreaming.

We spoke then about the power of the drums and I felt all the resonances of African magic. I have always known the drums spoke in more elaborate and formative ways than I could hope consciously to understand. As I listened to him I could hear in his voice how the cadence, tones and rhythms of the drums can themselves shape the way reality is modulated — how words can be entirely circumvented by the authority of rhythmic patterns.

Mark broke in, raising the issue of the Junkanoo, their annual procession held carnival-style, each Christmas. It has become debased, he says, merely a tourist Mardi Gras in which the participants have forgotten the true purpose of the street dance amid a flurry of prizes for colorful costumes. In spite of this unhappy state of affairs the magical goatskin drums are still used, igniting fire in the veins as they have for generations.

Mark wanted to restore the fundamental meaning to the procession but he also has a more ambitious plan. In the unexploited original, the dancers wore white and the whole affair had a marked sense of rightness. While he was talking, I flashed on my earlier perception of the dolphins helping co-form the rhythms of the material planes, within the limits of their domains, by naturally playful and collaborative behavior. Mark had said in a previous conversation he was aware of this concept and I realized as he conscripted David's participation in this new and righteous Junkanoo, this ceremony itself may have some infinitely more subtle functions. I saw the political savvy in asking a Haitian to put his vitality behind the drumming.

I was to find out later from a middle-aged cab driver the Haitians were largely considered the scapegoat for the island's woes, at least in the view of the older generation. The young Haitians he described as fearless, violent and without a care for anyone's safety. I was intrigued when I heard 'fearless' and thought of Brother David and his friends and the power inherent in redeemed voodoo.

Mark, by picking David as the main drummer for the Junkanoo was symbolically closing an ancient debate. As the story goes, the black races have traditionally found it hard to get on with one another, almost invariably splitting into small and mutually hostile groups. Some point to this being the reason a handful of Europeans could catch and enslave so many Africans under conditions which could only have been rather difficult for the Europeans: add to that, the cultural and spiritual beating the islanders in general have taken at the hands of the Spanish. Only recently have they started the long climb out of the violently difficult chaos engendered by such savage colonization.

Haiti is no exception; doubtless many of the young have been bent sideways by bitterness. Mark's decision however would go a long way towards uniting the spirit of the Bahamians with the occult abandon of the native Haitians.

After a final pipe our interchange felt complete. Amid the customary hand-shaking, slapping, fore-arm grasping, I found myself teetering backwards under the God-intoxicated gaze of yet another young rasta. I never did catch his name among all the mumbling, but I got a taste of the raw power many of these young sorcerers possess. My companion commented later on this healing touch, her arm tingling for a full half-hour after being shaken and Mark laughingly agreed as we bid our final goodbyes.

We ambled back into town; Mark seemed delighted by the encounters with Mr. Fussell and the young prophet. He offered to help us find a boat and someone with a fair knowledge of the waters. We were joking between each other about allowing the dolphins to 'find' us, since by now we had learned they are seldom where you think they are, when Mark broke in, to our astonishment, with "I can hear de dolphins talkin' Mon. If dey go under de boat I can hear what dey sayin' ".

I had no doubt he was telling us the truth. I told him my rastaman story, how each morning the old man swims straight out to sea . . . (Mark's eyes shine delighted), realizing, as I am talking, the absurdity of telling a head-honcho rasta saint what is probably island folk-lore, when BOOM, white inspiration, and I **know** that old rasta was transported back to shore, and probably in high style, **by the dolphins**! Then it dawned on me I had not been speaking and I saw the broadening smile of understanding on Mark's face. As we both telepathically reached this image, there was an electrical feeling of bonding — nobody spoke for a long spell before we broke up in the elevated joy of the moment, laughing and nodding hugely to ourselves.

We spent the day after meeting Mark exploring the perfect little tidal pools lodged in among the volcanic rocks of the Eastern coastline of Paradise Island. Time to pause in the sun and the microcosmic integrity of the pools.

We had reached a point of multiple choices. Options were coming out of our ears and both of us felt ourselves coaxed to extend our stay and follow whichever trail led most fluidly to the wild dolphins. By now we had ascertained they rarely come near the inner islands because of pollution, noise and the density of the water traffic. To have a chance to swim with them on this trip would mean going to Eleuthera or Grand Bahama and that was going to take a fast boat and a lot of good fortune. Mark or Henry might produce a boat but, given the slower rhythms of island life, it was frankly unlikely to happen within the week. There were, of course, small inter-island steamers, each, so we were told, adopted by a dolphin as they enter the various ports.

Then, as if to give an across-the-board field of possibilities, enter John and Karen, a nature-hungry American couple, offering us a ride in their thirty-five footer, if the weather proved amenable.

As we hung out by one of the pools, John related how on a number of occasions, he had been lying back at the helm of his boat, coasting along in a good breeze when, inches from his ear, would blast the sound of an express train going through a tunnel. A most tremendous explosion of air. After he had retrieved his heart from his medulla, he would look down and see this single, sleek eye, half-closed in humor and contentment, beaming up at him and effortlessly pacing the boat.

He told me dolphins expel air from their lungs at velocities of something over one hundred miles an hour. Scientists also tell us dolphins evacuate 95% of the air from their lungs, somewhat more than the 40% to 60% we humans are used to. Within the Yogic reality, breath holding and breath control in general are regarded as being the main portal to other levels of consciousness. With what we have been able to observe about the dolphins, might they not have stumbled on this natural technique for manipulating their consciousness?

Putting aside for a moment all these options

whirling in our heads, we settled down to make a detailed study of the largest and most beautiful of the tidal pools. We saw how the flow of water had shaped the rocks into the most sweetly edifying micro-environment. The ever-to-be-seen miracle of natural law and inter-connectedness carved small individual caves no larger than billiard balls and, on the balcony of each, sat a sea urchin colored in the deepest reds and black, the rock-face in each cave worn smooth by the small daily movements of the urchin.

Below the urchin condominium there sparkled an under-water forest with finger-long, transparent fish flicking between miniature redwoods. A small hillock, worn partly hollow by minute tidal waves and the buffeting of the ages, arose out of the forest to within a few inches of the shimmering surface like a onetime fortress, before the rock fell away in the deeps. Here no life attempted to cling to the smooth, curved sides, fancying in the way of all life everywhere, the shelter of more protected regions.

While I attended these miracles, the thought came to me I was looking at a beautifully sculpted model of a full-scale location for an inter-active environment within which humans and dolphins might live and play together.

My mind switched gear and I saw humans living in the cliff-dwellings on the high Western rock-face. I saw restaurants with glass-sided walls; private areas where dolphins and humans could swim and play together; pools in which women could give birth underwater with the help of dolphin midwives; cinemas in which dolphins could watch our movies (should they want to . . .)and we could, in turn, watch them; underwater houses for those wishing to live side-by-side with dolphins; soft and amorous scientific research units; places for lovers and poets and artists. All this erupted before my eyes in a macrophagic miracle, and I knew in my heart I would see such a place within the next decade.

Back to the options: After oscillating wildly over the various possible courses of action, it struck both of us that one of the great pulls of the islands lay in their lulling, beckoning allure. How many people come here, we had met some already, and for one reason or another never leave? We were teetering on just such a brink and the breakthrough came when we appreciated how our adventures will manifest day-after-day, wherever we are. The wild dolphins will wait for us.

Meanwhile we had some unfinished business with the Spiritual Nobility of the Island — and another glimpse of the potency of those within the **Network of Light**, what we have come to call the growing matrix of conscious planetary citizens.

If all plans were to fall into place, life would be a miserably predictable affair. Glimpses of the future are one thing, but too sweeping a panorama can easily tumble a 'sensitive' deep into a private hell of plane crashes, assassinations and similar catastrophes, not to mention the monotony of a life without surprises. As a race we seem to be able to predict enough to be useful but never enough to entirely preempt the experience. And perhaps planning is like that too. We can plot and plan something, but if we are fortunate the eventuation will always be more exciting, more surprising, and invariably more humorous, if at our own expense, than ever we could have hoped for in all the planning.

Saturday evening was one such occasion! Our final evening, and my ever-cycling old rational brain, sifting, watching, early-mammalian edifice of manipulation, must have created a picture of what it wanted to see. After all, Henry might be there with his guitar — I phoned him twice but had to leave messages — and Brother David might be able to get the magic drums for the night, and we all knew the power for change on the islands lay in the music . . .

Evening falling found us, therefore, laden down with guitars, micro-organs, tape-recorders, happily making our way into Mark's back garden.

To our astonishment, the entire place was transformed, house and garden; everything bright and cleaned, straightened, propped up, laid down. Both goat and kid looked slickly clean and wee half-AIO had the new confidence of a life lived all of two days. AIO herself must have settled her differences with the dogs because she held her head high and didn't stir while her offspring nuzzled our crouched knees with its softly-furred little slab of a wall-eyed face.

Mark was in the shop with half a dozen friends. I sensed they had been carefully chosen. Eli was there to our surprise, sporting a new pink T-shirt with SUN YOUR BUNS IN THE BAHAMAS emblazoned over the heart. I fell on his shoulder laughing with the miraculous synchronies of island life. My companion, the previous day, had slipped Eli five dollars to get himself a pink T-shirt — she felt the color would help him cool out his fascination with his scorpionic qualities. We knew he had hurried off to buy it but neither of us had seen him since.

Significantly, while looking through the dilapidated and forlorn Nassau shops, we had noticed and laughed at the SUN YOUR BUNS legend. It became for us something of a private joke which, in the deserted rocky pools of the morning, we had been able to put into practice. And here it was again!

Back in the shop, much hand-shaking, greetings, everybody a little shy at first. Mark however was open-hearted as usual and roaring in mutual laughter. The atmosphere relaxed tangibly as we sat together uncrumpling twists of brown paper or newsprint, and prepared our herb. The music started. Mark clapping and humming. Slim, from New York — "I came here for two weeks and never left . . ." (Ah! Ha! A small confirmation of our decision to return) — stabbed away at the miniature electronic organ; Eli played the base with his mouth; the others moved and swayed in the half-light of the 40 watt bulb.

Elizabeth, Mark's sister, lit some incense and the Spirit of Grace overshadowed the house. We talked about the previous night and agreed it felt like a psychic storm blew over the island. "De wind of change," Mark called it.

The evening accelerated in intensity. People came and went witnessing this most curious of gatherings. As the first group rose to leave, I related the events of "the bite . . . the bite . . ." Lizards had come up in the conversation; Greg, one of the Bahamian rastas, telling us how territorial they are. He was beside himself with excitement because a lizard had actually allowed him to sit on the same rock earlier that morning — apparently an almost unheard of phenomenon.

When I got to the place in the story in which the furies had broken loose, two of the dogs, hitherto sleeping peacefully, erupted monstrously into a yowling, screeching, gnashing fight. They stopped almost instantaneously and after the silence died away and amid a growing surge of relief-laughter, I continued the story to its fruitful end. I could see from the wide-eyes, future versions of this evening will include reference as to how the dogs were momentarily dibukked as an illustration of the white brother's tale. Of such things are legends made.

The feelings intensified. More exits, just leaving Mark, Eli and the two of us. No sign of Henry. After some talk about who should go and fetch him, Eli, trepidatiously, eased off to see whether he could extract his brother from a busy Saturday night's trading and invite him to a musical evening. An unlikely prospect by all accounts, but certainly worth trying for.

I should say at this stage, and writing in retrospect can see it somewhat clearer, I had become a trifle attenuated from among other things, enormous excitement, a certain lack of food, my mad-dog penchant for the Noonday sun, and needless to say, a great deal of excellent Rasta herb. It has been known, in that finely-tuned state, for a clear and high level of consciousness to turn on its bearer. Indeed, many of the difficult spaces in which one can find oneself with the noble entheogens, seem to be tripped off by an open consciousness suddenly getting spooked by someone else's fear.

In truth, I can now see that Eli must have been terrified by a further encounter with his brother. The fact we accompanied him, and were the first to do so, had evidently won us his ongoing gratitude, but the idea of taking his so recently healed relationship with his brother one step further really required a great deal of personal courage.

He left, anyway, although I have wondered since whether he actually did go and see his brother.

After Eli had gone, I hit the skids. Paranoia swept through me, un-nameable fears more subtle and beastly than any realization of the physical dangers of being alone in the middle of a volatile island 'ghetto.' Mark had re-tuned my guitar by this point and was singing his own profoundly spiritual rastafaree songs. His voice was deep and brown, melodic and confident of itself. The more beautifully he sang, the worse I felt.

I made no pretense of it, and perhaps as a consequence, in the middle of the hurricane came a moment of peace. I sensed with my inner-eye that Mark needed to see, feel and comprehend my vulnerability. Any residual male-macho qualities picked up from island life dropped off of him; any shards of confused respect for British stiff-upper-lipmanship fell away. It was a close moment between people from very different cultures.

Then the hurricane started its raging again. Mark sang a second song, a long and majestic roll-call of all the tribes of Israel. Somewhere in the middle of it all there was another explosion of barking and butting as AIO, who must have snuck unseen into the shop, and one of the dogs, rolled and slid on the dirt floor. Mark motioned me not to stop them. I felt a shudder sweep through me and in one golden moment knew the animals had taken the dibbuk and were running it out into the garden and off into the night. The affair ended within a few seconds; the fight over, and I felt once more in a state of righteousness.

I asked Mark to play his song again and this time the reverberations and rhythms coded into the Hebrew words filled my soul. I could **hear** the underlying meanings. The Spirit uplifted us and the little room filled with light and joy. Eli returned, falling into the calmness sans brother — Henry apparently really was too busy — and I saw why it had to be just us three on this special evening.

The talk ranged over many subjects. I broached the extra-terrestrials and Mark registered no surprise. I told him we both had the feeling of alienness from Prophet David, a sense of him coming from 'elsewhere.' Mark reminded us the young prophet is half-Bahamian and half-Haitian. "Dat's his inheritance. All de power of voodoo, given to him now to make good." He smiled again at the wonderful reversals. We went on to talk about The Urantia Book — as religious books go it is particularly significant since it maintains it is part of a new and continuing revelation. This interested Mark and we promised to send him a copy when we returned to New York. He was delighted, as the island, he told us, was starving for fresh information.

If I had any concerns as to Mark's reading skills such thoughts fast disappeared when he mentioned he had read the encyclopaedia. The entire encyclopaedia! And not just one, but **five** different versions! This one will enjoy the precision and density of The Urantia Book, I thought, and laughingly caught my companion's eye.

Sometime over the next few hours, when all had been said, and felt and seen, we rose — levitated it seemed — to go. We shook hands and held fore-arms, feeling the three of us deep within each other's hearts and souls. A connection had been made which will last an eternity. In the most mythological of my reveries, a meeting had taken place between the Spirits of Beauty, Goodness and Truth. And these Spirits overshadowed us mortals, to meet and converse in material form.

We all three of us felt profoundly privileged.

After a full and deep sleep, some more explorations of our tidal pools, this time in the face of a rising Westerly storm, great breakers falling like mountains on our urchin villages, and oddly, not stopping to say our goodbyes to the dolphins, we slipped fluidly back to the Big Apple.

"What! couldst thou not perceive that at the entrance to all the grander worlds dwell the race that intimidate and awe?"

Zanoni: by Edward Bulwer Lytton
1837

29/9/00

CHAPTER SEVEN

I approach the next stage of this journey with a little trepidation, for it concerns one of the most shocking thoughts available to any sentient species. I have wondered now for three years whether this fearful idea needs to be aired. "What a race doesn't know, doesn't need to be told," I have attempted to persuade myself, but that is too simplistic a concept, as well as being condescending.

Yet now, and only now, do I feel it a necessary concept to explore since, shocking though it might be, it contains also all the glory and compassion so characteristic of the best of the human spirit. It is ultimately a story of victory and redemption. I am saying this in advance, since what I was to discover can be summed up quite straightforwardly as 'the worst is over.' We as a sentient race are at last emerging from the darkest of dark ages into a Universe as benign and fascinating as we can conceive. Though the mopping up, in its many and varied details, is by no means over, the Victory is won and the immediate rewards of any movement towards the good are, and will be, bountiful.

With that said, I can now embark on what will, I'm sure, turn out to be only one small holographic shard of the macrocosm. But if this shard is perceived as having integrity, it should contain all the major elements of the drama being enacted now on this planet of ours.

So, what is this thought which is so appalling I have wondered whether even to utter it?

It is simply this: what if our planet, our race, you and I, have been, through the ages, either the battleground, or worse, the playthings, of another race from another dimension or another planet? What if all our strivings towards Goodness, Truth and Beauty have been systematically destroyed by entities so cynical and brutal that they have thought nothing of the worst type of colonization?

And there is worse: what if this race has had intentions so lowly that other, more benign observers, have been hesitant even to allow the true horror of our situation to emerge?

And worse still: what if this malignant race has been controlling the very consciousnesses of our planet? Sitting, as it were, just the other side of the range of our normal awareness, waiting to torture or mutilate any of those brave enough to step outside the rigid confines of conventional thought? What if the first entities met in our exploration of the "invisible world" have been fearful and malicious?

What if the War in Heaven, mentioned with unpleasant regularity in most of the myths and histories of the planets long past, was a **fact** of Universe affairs? And this planet, not perhaps the one most centrally involved with the frightful depradations, but involved none the less with much of the unpleasant side-effects, and through little fault of its own?

If indeed this were all true then what hopeless futility would have been our lot! It would be as if we had nowhere to turn. Our every effort to transcend, blocked.

To understand, torn down and vilified by the malignant ones and hidden from us for our own good and peace of mind, granted, by those who wish us well. Caught irrevocably between a rock and a hard place.

As I said, it **is** a thought — history will no doubt confirm or deny the actual reality of the situation — and it was as a thought I was being required to confront it. This is important to stress since in all **my** travelling through the inner and outer realms I have never had to deal with a truly implacable and merciless enemy. Possibly others have, and God bless them for taking the brunt of it, but then again, maybe the totality of unredemptive evil has always been merely an idea. A product of our imaginations. A potential perhaps, but never anything with any lasting **reality.**

At the time of the experiences I am about to relate, however, it was all very real. The Guardian of the Threshold always is. Whether it was the snake who entered my eye or now this, the War in Heaven. I **needed** to perceive those entities concerned with utter reality or the significance of their messages became meaningless.

What, after all, do we know about the proverbial War in Heaven? The Zorastrian cosmology, probably the oldest one to which we still have access, speaks of two warring brothers. The Sumerian Epic of Gilgamesh tones it down somewhat but still speaks of Huwawa and the evident existence of entities with little vested interest in human betterment. The Egyptian Set is a thoroughly unpleasant piece of work. Nordic and Greek myths talk of Gods and Goddesses, vain, malicious and power-hungry; almost everywhere we look into the old stories we see this pervading cynical element, invariably seeking to pull the good down.

The Gnostics, once suppressed by the Church

Fathers, are again making their vision available to us through the recent archaeological finds at Nag Nammadi. A depressing vision it is too. A world created by a demiurge in opposition to the better judgment of older and wiser Universe Personalities, designed to trap and seal off the one thing we, as human beings, hold dearest — the presence in our hearts of a Divine Spark.

We have spoken of original sin in our Judeo-Christian Western cosmologies as if it were indeed the status quo. As if we have been inevitably fated to be born into an evil and degraded world, any effort toward redemption bound by entropy to collapse once again into chaos.

We have lived, if we could but see it, in a depressingly pessimistic age and perhaps there has been good reason for it. The more coherent of our recent cosmological analyses capture our thoroughly ambivalent situation with what is likely to be seen in the future as a **revealed** perception. Doris Lessing would probably dismiss her *Canopus in Argos Quartet* as a work of pure imagination, and yet has the grace to admit in her preface to Volume 2 that she has "little idea where thoughts actually come from."

The picture she paints is intuitively accurate however, even if the main protagonists may not be those she names. She too sees this planet as the theater of three dominant colonizing forces each drawn to new dimensions by their mutual interaction.

John Milton, another contemporary, seen in the larger perspective of Universe affairs, also introduces the impact of the downfall of an angelic entity on the life of the planet. His vision, of necessity, is couched in the more traditional terms of Christianity, yet tells a story which is echoed in virtually all cosmologies.

Thus an honest appraisal of this planet's history

from the broader vision of a populous and organized Universe, must admit the presence of some very unhappy elements. That appears to be unarguable. But where can we look to see a more comprehensive picture of this difficult issue? It has appeared so shocking to thinking people throughout history, I can only put together the pieces which have been allowed through and map them over the particular cosmology I regard as nearest the Truth.

This brings me back to The Urantia Book, which I had started reading when first becoming involved with the dolphins. A document such as this, I was discovering, stands or falls on the inner coherence of what it tells us. There is little or no way of proving it at this point in time, although it contains many references to objectively verifiable factors. Our scientists are unlikely to leap at the chance as yet.

And this was precisely what was starting to happen in my life. The Urantia document, which claims to be dictated by entities from the celestial and angelic realms, was turning out to be the central piece of information I needed to resolve the warring elements in my own mind.

In a direct way I could see how my introduction, first to the dolphins as a sophisticated and advanced culture, and then, the encounter with the young boy in Bennett Park, prepared me to accept the **reality** of the larger Universe domain.

Whether the dolphins actually turn out to have all the wonderful qualities I perceived in them can only be known through personal involvement and future biological discoveries. Regardless, what they had allowed me to do was to permit myself the **reality** of sharing this planet with another sentient species.

The incident with the flying disc and the young

boy had taken this a step further and introduced me to the **reality** of a populated Universe.

Meanwhile, The Urantia Book, with all its overtones of Mid-Western conservativism and incessant bureaucratic meanderings, was starting to give me a Universal **context** within which to make some sort of coherent sense of those experiences. Although, at the time, it didn't feel like that!

Having become acclimated to the Urantia vision and allowing its truth and coherence to permeate my own thinking, I was suddenly confronted with what the writers term the Lucifer Rebellion. What was particularly upsetting about this Rebellion was not so much the fact of it, since there's still much to know and understand of the protagonists' viewpoints, but the bloodthirsty attitudes of those reporting it. It appeared from what they said that they could not wait for the moment when Lucifer and Satan were to be exterminated.

Later I was to come to see the very gloating tone of those writing about the Rebellion and 'termination' of the incorrigibles, was in itself a clear signal. Either the reader became sucked into the rhetoric and emerged a glassy-eyed cultist, or he made the necessary leap of intelligence to see there was something suspiciously wrong with the vitriolic relish employed in certain passages.

As I entered into that dark night of the soul however, and the winter snows of New York yellowed the skies, it was with an intuition of unforeseen changes to come, amid what I can only describe as a mental and emotional hurricane.

I was living The Urantia Book with a passion, having decided to accept its truthful content. I'd read its fourth part first, a description of The Life of Christ by the angels who accompanied him and had witnessed the events. It all struck me as wonderfully true and I felt it directly in my heart.

At Easter I found myself reading the passages concerning Jesus Christ's death and resurrection and **knew** from the flooding emotions that indeed a God had died on this planet. In those moments I met Christ Michael, the Creator Son of this Universe; I met him unequivocally in the depths of my soul.

It was, as I said, all the more shocking then, to arrive, subsequent to this revelation of immanence, at the story of the Rebellion and the harsh tones in which it was viewed.

The story itself is a straightforward affair, in as much as any such description of wide-reaching cosmic affairs can be straightforward. It gelled too with the scant available information garnered from other cosmologies, but in a manner that threaded together the personalities involved in an integrated and comprehensible pattern.

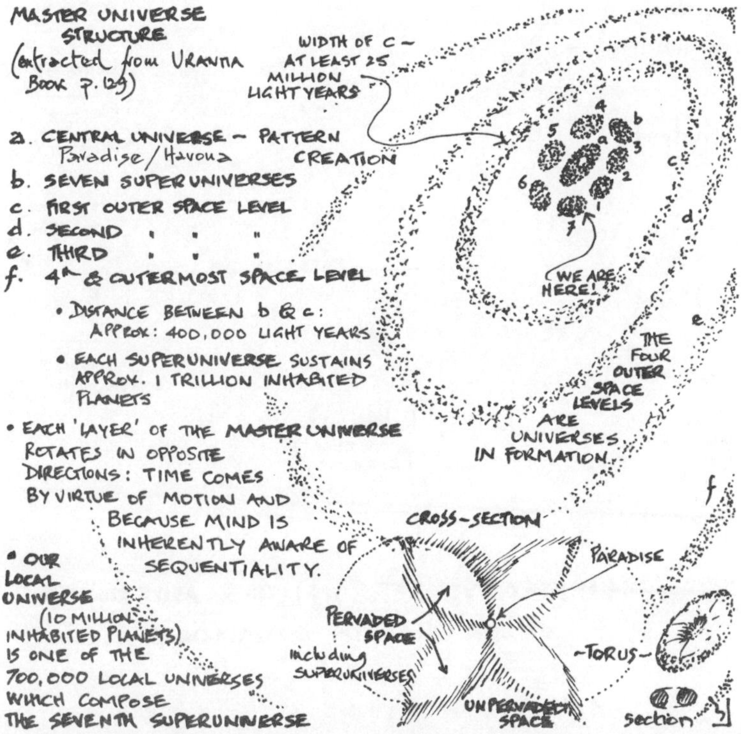

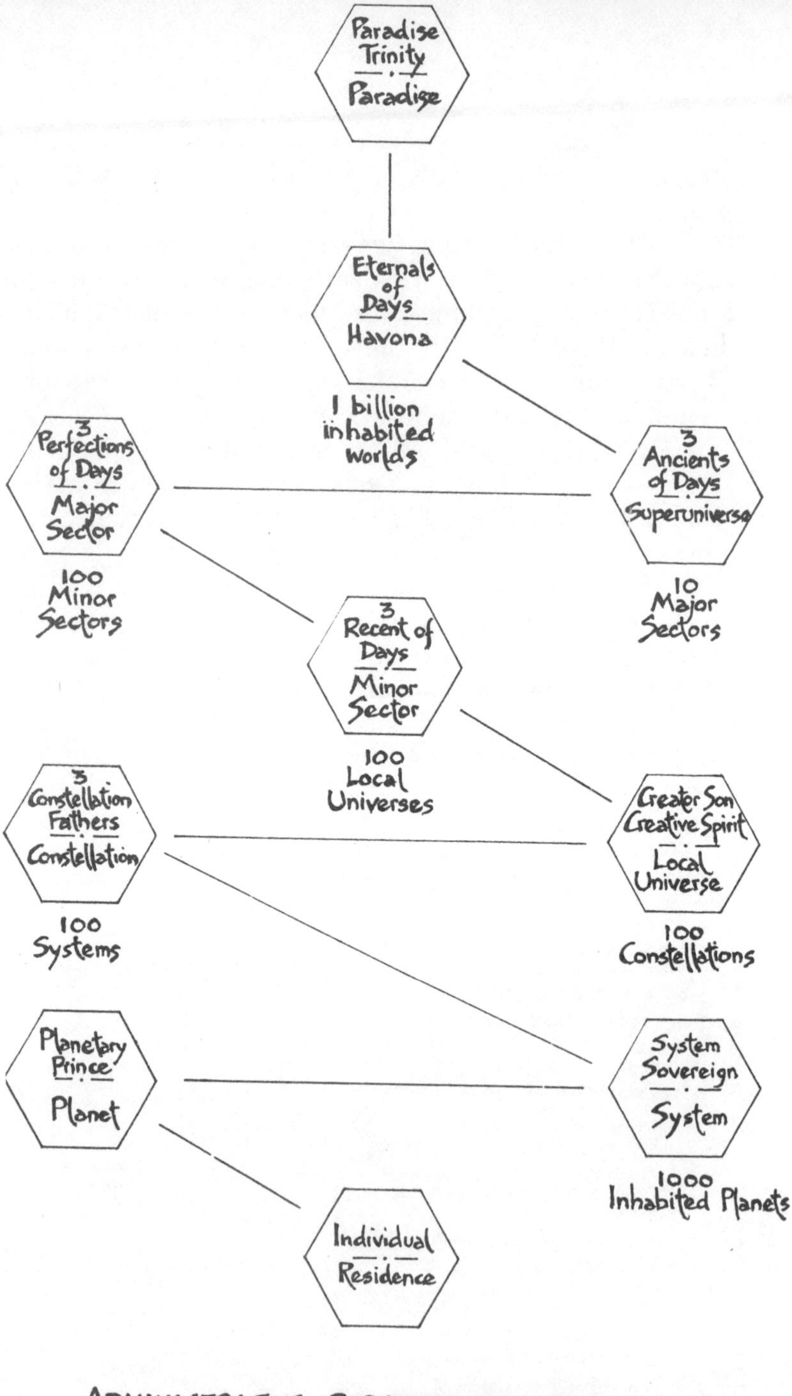

ADMINISTRATIVE PERSONALITIES AND UNITS OF THE URANTIA COSMOLOGY

Lucifer, so the angelic communicants tell us, was, at the time of the Rebellion, which they place as occurring a little over 200,000 of our years ago, System Sovereign of our little area of the Universe. This means, in the administrative ranking of celestial hierarchies, he was the angelic personality given charge of the one thousand inhabitable planets within our local system. Although, at the onset of the Rebellion, only six hundred and seven of these thousand planets were populated, our sphere Urantia was numbered 606.

Lucifer and his main assistant, Satan, were responsible to the Universe Administration for the growth and prosperity of these worlds, but most of all for what is called the Mortal Ascension Plan. This plan, known about and acknowledged on more 'normal' worlds, ensures that mortals — human beings — are guided through the University of Life by their angelic helpers and, on dying, up through the Mansion Worlds to System Headquarters and beyond. It was a substantial responsibility, considering the billion upon billions of individual souls in their charge, but within the enormous aegis of a Local Universe, a relatively junior post. Not that Lucifer and Satan were not created for the function. They were, we are told, despite their rather individualistic slant, both selected from large numbers of potential candidates.

If all this sounds a little plebeian, take a moment to seriously consider what an inhabited Universe means. Regardless of the state of advancement of a planetary culture, there is bound to be some form of Government, even if it is merely a co-ordinating agency. Enlightened self-government may well be an aim to which all aspire, but in the interim someone has to supervise the communication links and fix the drains.

And unfortunately, when you find an

administering body of officials, you also tend to find a bureaucracy. Indeed the endless list of hierarchies and pencil-sucking precision of much of The Urantia Book itself speaks very convincingly of a Universe almost completely snowed under by bureaucratic tedium.

Perhaps it was this very factor which so incensed Lucifer and his colleagues, after many years of deliberation, to declare open Rebellion against the Cosmic Authorities. Possibly it was not at the time even considered rebellious by the instigators, but merely an expression of their own dissatisfactions.

No administration, however benevolent, welcomes disagreement, and the one that confronted Lucifer showed every sign of wanting, for reasons of its own, to polarize the protagonists rather than seek reconciliation.

The actual issues involved are not the major question here, the only description of them being from the Administration itself and therefore undoubtedly biased. What was more important was the way in which the Rebellion was dealt with.

We are told the Administration banner was carried by Gabriel, not the Archangel of Christian cosmology, but the first-born of the Sons created to run the Local Universe. This puts him in overall charge of no less than 10 million inhabited planets, a very substantial hunk of real estate.

Having made his decision and on issuing his proclamation, Lucifer, by all accounts, was not directly confronted. Instead the opportunity was used by the Universe Authorities as one of those situations of moral choice which provide sentient creatures with the chance to make quantum leaps of wisdom.

All the entities within the local system were given the option of following Lucifer and the rebels in their call for greater freedom of choice and self-government for all orders of intelligence. Gabriel is mentioned as being particularly instrumental in developing the options. In spite of his staunch rebuttal of Lucifer's claims, vast numbers of angels and some thirty seven planetary administrations decided to follow their System Sovereign's lead. Urantia, our planet, was one of those thirty seven.

It is necessary for a full understanding of this situation to know a little more about the way an individual planet is regarded within Universal Affairs.

Although it may come as a shock to some, and there's a very good reason as to why it might, planets and their animal populations do not appear by some freak chance. The evolutionary theories of Charles Darwin, and the more recent essays of the sociobiologists, merely outline how life develops but make no coherent attempt to suggest how it got here in the first place. The 'primaeval soup' theories of some modern biologists and astrophysicists are simplistic and parochial in the extreme. Those who have studied the early years of life on the planet with an open mind have invariably detected an intelligent hand at work.

And so it was. The planet was created by processes currently unknowable by us and then seeded with life by Beings whose range of influence covers the whole of pervaded space. In a Universe of trillions upon trillions of inhabited worlds, this apparently is no big deal.

Once seeded, life processes are nurtured in such a way as to produce a creature capable of will decisions, and ultimately, knowledge of the Beings who created them in the first place. As you can see, the span of those activities is truly vast in both spacial and temporal terms.

When, after the countless millennia pass, a creature does emerge, develop and show signs of will and moral choice, then the Universe Authorities deposit a small group of, what we can only term, extra-terrestrial personalities, on the face of the planet. In this world's case, such an arrival was seen approximately 500,000 years ago and among the personalities involved was an entity called Caligastia with a staff of one hundred. Quite how this occurred will have to wait till later in the story but suffice it to say Caligastia was, to the Universe Authorities anyway, titular Prince of Urantia.

It was therefore Caligastia and the majority of his staff who made the decision, 300,000 years after coming to this planet, to join Lucifer and Satan in the System uprising.

Upon Lucifer's Declaration of Independence, the Universe Authorities promptly isolated this System from the rest of the Universe by cutting off access to communication links and certain sustaining cosmic rays. All relevant Universe personnel, having made their decision whom to follow, were left to their own devices within this sector of the Local Universe. Lucifer had in fact got what he wanted although not, I suspect, in the manner he might have wished. If the state of this one planet has been anything to go by, then few would agee that his policy of self-determination really worked out to the best for all.

Indeed this story is no apologia for Lucifer, his actions or his decisions, however much I may have sympathized with his initial intent. He bit off far more than he could ever have chewed and the results of his action have been obvious.

No, something more was going on. Here was The Urantia Book, for me vehicle of such wondrous revelation, the reading of which had allowed me finally an intimate access to Christ-Michael, the Creator Son of our Local Universe, suggesting, encouraging even, the support of thoroughly unforgiving and strident attitudes of rebuke.

Surely, I thought, the entities who fermented the rebellion have suffered enough by now. The evident failure of their plans for premature freedom of choice must be very obvious to them. However insane, however irresponsible, however frightful the consequences of their actions, no entity in a benign Universe can deserve ultimate termination. What's the use of that, I wondered? All that valuable experience down the drain. And anyway, the whole affair may well have been built on misunderstandings. Lord, I've been thought insane or irresponsible in my time by enough well-meaning people to know how easily such confused perceptions arise. Something else was going on and I knew it in my heart. I remembered the encounter with the young boy and his look when he told us of those who take on the deeper problems; of them becoming the vote carriers of their races; and of a democratic, one-person one-vote system.

I wondered how many people were being, or had been, requested to take on **this** issue. Both of us, my companion and I, knew Lucifer and the others could not be terminated while there was any remaining sympathy for them.

And taking the facts as we'd been given them, By God, we sympathized!

We agreed that if we do have any voting privileges, and given the nature of the odd occurrences, why not, then there had to be a better way. A resolution in which everybody won, all Beings came out ahead. A reconciliation in which Lucifer and Satan's courage to defy an entire Universe was recognized, and all they had learnt in this long and painful detour made available. In **our** Universe anyway we wished to see these rebels redeemed and all Sons of God working together again in collaboration and cooperation.

I have given only the briefest outline of a wearying and tumultous period as Spring turned to Summer. Dwelling on these issues and considering the horrific possibilities of living on a planet which, through little apparent fault of it is many inhabitants, had racked up so many millennia of distress, brought in its own particular brand of misery.

Yet the decisions we had come to were clear and unassailable. From time to time we both thought we'd gone off the deep end, yet these were the real factors in our lives and all we'd studied and lived through led irrevocably to these moments. We wondered whatever those guiding us would now show us. Would our thoroughly human decisions be taken into account? Would we be shown by ensuing events that our request for a Reconciliation had been heard?

The news was not long in coming. And, in the way of the subtle nature of these Universe affairs, needed a degree of faith and trust in order to perceive their connected significance.

We heard of some strange events in Canada. A man whom I'd not met or talked to and whom my companion knew slightly from some years back but was not in current contact with, had had a most surprising and startling vision.

First let it be said that Nicholas is far from a visionary in any conventional sense. As a personality he is down-to-earth and extremely well-grounded. His Italian extraction and Roman Catholic upbringing served to give him a religious orientation, but the responsibility of four children and a thriving engineering company meant a predictable drift from the tenets of the mother church. Somewhere along the line he too had encountered The Urantia Book and had been an avid reader for some years when he had his vision.

The small community of people who read this book are a loose-knit network in a number of different countries but focussed mainly in North America. The co-ordinating agency is based in Chicago, where the original revelation was received, and are, though pleasant people, of a somewhat conservative nature. They see their job as the dissemination of this book and are concerned more with preserving the sanctity of the Revelation than in spreading a new religion. As well they should. The whole point about the information in the book is that it is of a thoroughly personal nature. What is important is for an individual to receive and apply the insights **without** the resultant burden of an intercessory priest-caste. That is indeed the main reason why it was given to us in such a covert fashion, so as not to create another personality cult.

Well, the Chicago Foundation could never be accused of personality cultism even if they do fall into some of the bureaucratic pitfalls of almost all human institutions.

This background gives a context within which to appreciate Nicholas' vision. Whether or not he ever really swallowed the party line about Lucifer, in the way the rank and file readership of The Urantia Book appeared to have done, can be inferred from his reaction to the vision he was given. He is a good man with a fine heart and the true messages of the parable of The Prodigal Son did not, I suspect, escape him. He had also in his life, throughout the preceding year, to deal with the problems of an extremely difficult 14-year old son, so had the chance to evaluate the many ramifications of 'disobedience' within the confines of his family unit.

Possibly this challenge more than any other led him to follow the call of forgiveness when he saw the face of Caligastia at his window late one evening while visiting some friends in Canada.

Caligastia, you'll remember, is the deposed Planetary Prince of this world, left to roam the face of the planet through the long millennia causing, if we are to understand The Urantia Book's assertions, mayhem and general chaos. It was he, more than Lucifer and Satan, who had far more wide-spread roles to play, who is most appropriately identified with the 'devil.'

So here we have Nicholas one Summer evening, having retired to bed around midnight, gazing abstractedly out of a second floor window, when there, in front of him, peering in through the glass, was Caligastia, the monstrous devil of traditional theologies. It was not, according to Nicholas' later statements, a pretty sight.

Nicholas however, brave man that he is, summoned all the courage in his heart and, invoking the Spirit of Michael, advanced and embraced the shade of the deposed Planetary Prince. In that moment of cosmic intimacy, he told us later, he felt a tangible wave of forgiveness and love whirl through the Universe, not stopping until every rebel personality had been welcomed back into the fold of the Master Plan.

The power and the Truth of the vision was well nigh shattering for Nicholas, leaving him trembling in wonderment for weeks after it happened. This aftermath was also exacerbated by a total rejection of its implications by almost all his friends in the Urantia network, who wrote this revelation off as delusion.

My companion and I however, were not so hasty to dismiss it. Not only did the revelation itself have all the signs of a genuine experience, but the content of it directly corroborated the revelations and inner searching we had been going through all that winter.

It seemed to us to have the imprimatur of a far larger plan at work, as if an interlocking process of revealed knowledge was being experienced by a small group of people who, by comparing notes after the events, would be able to build an understanding of new developments in higher reaches of universe affairs. How else, after all, would these realities be communicated to us? We were scarcely liable to come across them in the media!

We had not, at any time prior to these events, talked with Nicholas or communicated our recent thoughts about a Reconciliation of the Rebellion. Neither had either of us heard of such a strange and powerful vision from anybody else at any other time in our lives. It came entirely out of the blue and seemed to dovetail so significantly into what we were starting to perceive as a major change in Universe affairs, that we could only wonder whether something which seemed so personal to us carried implications far wider than we could have dreamed. Our faith told us this was so, but our rational minds were still far from accepting it.

It was with this ambivalence that we both decided to go up to Canada to visit the house in which Nicholas had his vision. Our mutual friends in Toronto then hit us with the next remarkable piece of news. One of them, a young man who prefers to be called Edward, had started spontaneously falling into a light trance. When in this state, entities calling themselves angels spoke through him.

Our friends sent down a couple of transcripts for us to look at before we made the trip up to Canada and these too carried an unassailable feeling of Truth. They were of general interest and information and, although neither of us had hitherto put much faith in the words of trance mediums, those papers had a quality of integrity we recognized as coming from similar sources that channelled the Urantia Revelation.

Life was looking up again and we were soon off on this new trail.

CHAPTER EIGHT

"As you journey, you would be wise to see that there are ebbs and flows in the manifestation. There are whorls of energy that would cause greater alignments within the larger network of the planet. As many channels have opened, and as much information does come down, there is need for lateral co-ordination within the consciousness of the race."

This was an angel speaking. Our friend Edward lay back in deep meditation, allowing his body, his vocal chords and his neural wiring to be given over to angelic communicants.

The voice continued. It was making a great deal of sense, directly addressing itself, without any prompting to the pressing problems of impatience. Mainly mine, I suspect. Even Mark, the rastafarian holyman, had laughingly shaken his great head and rumbled on about going too fast. Within a few hours of arriving in Canada, however, we were swept into yet another adventure — contact with the angelic realms.

"Time is required for this and balance is of the utmost importance, in that the balance would be maintained between the information coming from higher and the dissemination of it within your own plane.

"This then would be the purpose of the networks,

to afford the flow of information so that many would be involved in the sharing of it. It would foster the Greater Understanding and, in the Vision, bring together the larger vessel which would be formed to contain that which will favor the next step. All members of the network would then be joined to work on receiving a greater understanding, thereby enlarging the Spirit capacity for same. Further, in this act of sharing is Spirit revealed as loving action. Therefore you are demonstrating your activity with Spirit, your connections with the larger frame.

"It has been shown to many in glimpses, the vastness of the nature of this network. Therefore, you would do well to contain within your personal and collective understandings, the connections aligning and forming a continuous link from this world and dimension, to the center of all things . . ."

As I write, some time later, I am calmer about the reality of angels talking directly to us. By this point I have done my reality-testing to my satisfaction and have no doubt such entities exist and are intimately involved with our destinies. However, knowledge tried and tested, and faith, are two different things. As I sat there, that afternoon in Toronto, my rational mind was in a turmoil. I could feel and sense the presences filling the room but the old left-brain was gnawing at the edges of reality.

Anyway, the effect came and went. Had I been more on the ball I might have followed up the angels' point about continuous links between this world and the "center of all things." It sounds like a direct line to Paradise!

However, instead I asked why we were drawn together, this particular group at this particular time. The angelic communicant, apparently not one for such obvious questions, replied: "Surely, then, it is love that has drawn you together." And then, more significantly, "This being the case then all is opened to be shared with you through such a channel. Abide then . . ."

Another voice layered in, somewhat different, slower and more ponderous. Edward would stir in his trance every once in a while, responding to who knows what. As every sensitive, he had battled through his own doubts. Modern psychology had been little help. Allowing himself to be convinced the voices were internally generated had deposited him very near the edge of madness on at least three occasions. Now he had become more balanced, finding an outlet for this extraordinary capability.

The second voice said: "I greet you. I am Shelenea. Being second-most attuned to the mind of this one, I come forth. My purpose is to communicate to you the essence of the healing in which you all share. The ending of the Rebellion is a healing. The closing of a wound is the step towards wholeness that the world seeks to know of."

We practically fell off our seats. The ending of the Rebellion! Suddenly something which had been of such consuming interest to both of us was issuing out of a voice purporting to come from angelic realms. It was unprovoked by us and was mentioned with such casual certitude we could barely believe it was happening.

The ending of the Rebellion is a healing the whole world seeks to know about, she had said, and indeed I was gradually starting to see the implications of this cosmic event. An ending to the disastrous situation which has been teetering on now for the last two hundred thousand years, obscuring our vision of this planet's place in a teeming and populated Universe. And, of course, with the ending of Rebellion would inevitably come the restitution of normal relations with a larger Universe framework from whom we had been cut off and isolated for all these thousands of years.

If this Reconciliation about which the angel was talking had taken place then perhaps the changes we are starting to see within our generations' view of extra-terrestrial life forms might be understood as a very significant move towards a restoration of normalcy. Only forty years ago the radio production of *War of the Worlds* sent hundreds of people panicking into the streets. The 1950's and 60's were full of alien body-snatchers, giant ants and robot killers until, with the exception of the hideous *"ALIEN"* designed by Swiss artist Giger, almost all contemporary visions have become distinctly more amiable.

The quantum leap in between the consciousness that produced the *"War of the Worlds"* and that of *"E.T. THE EXTRATERRESTRIAL"* and the popularity and effects of each, certainly indicate the moves we are making as a race towards opening our minds to the Cosmic Community. The earthborn fears of invasion from 'out there' are giving way to a signally more mature concept of otherness.

The angel was continuing:

"The effects then, of this healing in understanding will manifest in material form. In a greater sense of well-being, an unfolding of joy, a release of love long held in abeyance, a discovery of creativity, of fullness of understanding and the realization of the Peace Profound.

"This being the nature of the work in progress, that when these factors are more fully understood and claimed within your personal mind-grasp and that you would then **be** these things, the greater Being that emerges among you will be more fully appreciated and known."

Shelenea's voice drew to a natural close. Edward breathed deeply two or three times. I thought he was about to come out of the trance but another voice introduced itself.

"Greetings. I join you now as one who would be known as an Angel of the Future, carrying forward the visions which would be consolidated by the Spirit into the new realities you would form in your midst.

"It is within the mandate of our work to cull and cultivate the gems within the minds of humans who work for the advance of the culture and your world. We see with much rejoicing the work ahead of those who gather here for it is the release that you shall discover, with the feeling you will be coming into your own estate as co-creators.

"See the joy of utilizing the new forms of expression in sharing the news that you carry, for it is the New Gospel of **the release from fear.** It is the greater way of celebration that will uplift many into the more active expression of living with Spirit in all aspects of life.

"We would encourage you in your hopefulness, of your sharing of that which you hope for in your lives. These are the thought-essences which we garner, which we foster and which we amplify from the higher. You may see we carry within ourselves patterns from afar, which may be aligned with those thought-forms emerging from inner connections from the same source; these put together can manifest as new forms, the energy coming from the God we worship in unison. This then would be the revelation to our Orders of Being, of the Supreme harmonics which are invoked here.

"We then encourage you to be daring; we ask you to extend your grasps; we would affirm to you that you contain more than you feel today is possible. What we encourage you to do is grow in Spirit, become mighty in

Spirit. Then the greater works which are held in the Vision, but in the future as you would reckon time-unfolding, will be brought the nearer to you in your understanding."

Reading now the transcript of these transmissions, it is amusing and somewhat chastening to see the level of incoherence of our rational minds. The air had been so full of presences, the data so all-encompassing and joyful, that at those times the angels would stop talking and invite questions, I for one, could barely find my tongue. Thus it was in a rather hamhanded manner I tried to ask whether there was any way our interaction with the angels could be rendered more complete . . .

The Angel of the Future replied, apparently unphased by my confusion.

"The opening into this (state of greater completeness) is the acceptance of the possibility of the same; that it **is** possible is a great step towards realization. This you are making; this you are accepting and you are rejoicing in this.

"Be aware, friends in the flesh, the wisdom of the Spirit acts as the balance of all things in time; be aware that in your relative imperfection only that which you may do in safety and security, protected as you grow, will be allowed in the loving benevolence of the Overcontrol of the Universe.

"You are aware within your own minds of the quality of impatience. We see this in the positive light of exuberance of the spirit that is youthful within you. This we celebrate, this we foster. Yet, at the same time, it is the wisdom of the Inner Divinity that moderates in the co-ordination of the whole unfolding. The resolution of the problem for your souls is to find the stillness, the peace that God offers. This will quiet the impatience and offer the joy of serenity wherein may be safely contained the greater vision you seek."

So, the impatience issue again. But this time I was able to hear the deeper timbre of the answer and now, thinking back over the intervening months, I realize I have not been experiencing any of my old level of restlessness. With the reality of the Reconciliation growing firmer in my mind and the evident, though still largely hidden, ramifications of this healing on our world, I have had the increasing sense that everything is going well according to schedule.

The angel opened the floor to questions. One of our colleagues asked about the merits of LSD as a tool for furthering the Vision. The subject of entheogens, power plants which draw out the Inner Divinity, had come up earlier in the day and it was manifestly on her mind. It was an extremely pertinent question since many of us have found in our experience that the wise, shamanic use of these substances can substantially further the opening of the eyes of the spirit. It was going to be most interesting to hear the angelic perspective on this much maligned matter.

"Insofar as you have agreed to perceptual limitations and proscriptions as to the nature of your mind and awareness, these tools come to you as aids in the broadening of your understanding of the nature of your own being in its relationship with the Universe of Spirit. You have also been instructed as to the determinations of the usage of these aids and the proclivity which turns an aid into a crutch. Determination here is within the field of your own will.

"Aspects that will be realized will be the perceptions of fear or the realizations of love, whichever you choose. Given an assurance of the perfect protection of heaven, you would understand it would be an impossibility to lose the mind. It is better to see there is an opportunity to **use** the mind. In the fullness of your experience, you will discover that which is useful to you and that which is less than useful; that essential to your work and that which is non-essential.

"Insofar as this is the arena of your own choice, you then make the determination and the action is taken."

With this succinct and helpful statement the Angel of the Future stepped aside. I was very cheered by the evident lack of prudishness shown towards the entheogens, together with the assurance of 'the perfect protection of heaven.'

My companion then took the opportunity to ask Durandior, who had announced herself through Edward, about the dolphins and whales. She also included our Space Brethren in the question.

"Your question is a timely one," Durandior replied, "for it is Beings who would represent them, intelligences such as these minded ones, and yet not human ones, who would seek to communicate with you. We then shall leave the response to that question, for it is forthcoming.

"Know there are minds afar that would be keen in the study of the affairs unfolding on this world. There are those more advanced mortals on worlds settled in Light and Life that are most interested and concerned, insofar as they are manifesting the Supreme Deity among themselves and seek also to know of His Being in His Master Universe.

"This then is a point of confusion as to the nature of these contacts being made. In as much as physical laws cannot be transcended by physical beings, you would be wise to see that it is through the very Supreme dimension of Mind these contacts can be afforded. Entertain then, within your creative consciousnesses, that these manifestations are of an inner nature, projected by the mind into the external.

"As isolation may be maintained for a time, planetary revelations of the nature of extra-terrestrial/mortal contact must be entertained within the mind at first. Then the way is prepared in the future for the physical contact. Yet the communications, the intelligence and the love that will be shared will certainly precede this. Know that your technology does advance to a state where this is soon to be possible. This would be a contribution of science to revelatory religion, for the revelation of extra-terrestrial intelligence, that is both loving and understanding, will cause much to be realigned in the thoughts of many."

A long pause. Deep breaths. Then another voice, softer and more fluid.

"I am Talantia, Angel of Progress, assigned to the workings of this project." It was more than fascinating to think that a specific angel had been mustered for whatever she meant by 'this project.'

She continued:

"In the transformations that abound there is the co-ordination on a collateral level with other life forms on this world. You would be wise to look to the areas of administration where error is present, where sophistry has prevailed, to see where the greatest changes will be wrought.

"It is our concern that a harmony be brought forth within the consciousness of man and that of the other life forms with which he cohabits his world. The concept of 'dominion over' should then be revisioned. We suggest you work more fully with the concept of 'stewardship,' for in this distinction much will be made known to you of the nature of your tasks. In dominion, or domination, much fear has been generated. In the consciousness of the steward is the loving, gentle nature of the Father revealed to all. In this, you are called to do nothing less than to reveal the Father. The Father is revealed through you to your brethren but also to all life that conjoins with you in the realization of the vision of peace.

"Know there are vast potential areas of co-operation within the area of the planet that is aqueous. Herein lies much held in reserve for the future of the race. Our plan is to bring together life forms for the greater pleasure of all so that it will be with delight and joy you will find union with intelligences not of your own nature. In conjoining and cooperating with these you will gain insight **on this world** of the nature of the higher worlds where greater varieties of differentiated intelligence function in a Supreme Harmony. This then is the purpose for these loving intelligences to be among you.

"Look to the whale, the porpoise, the elephant, the horse; look to those creatures that are close to you in your evolutionary development to find areas where you will meet in joyous cooperation. Beyond that, you will also know the power and presence behind the other forms of life on this planet. Look to the power of the great trees; look then to the healing of the food plants; look to the realization of all these seen in joy of the life flowing through them from the One Source. When man comes to the fullest understanding of this revelation then will heaven be descended upon the earth."

And so ended the first session with the angels.

Contact with the angels and the profound sense of exhilaration we were receiving from being in the presence of more finely tuned intelligences, were having their effect. The sensitization of the session carried over into our lives and in a gentle but persistent way, upstepped all our consciousnesses.

My thoughts were still very much concerned with Lucifer. I knew by this time I had 'taken on' the identity for a period; that I had volunteered for the role for who knows what true reason. The pride issue, predictably enough, reared its head. I found myself oscillating between my own feelings, ambivalent in the extreme — how could I, mere mortal of the realm, a piece of driftwood in the Space of Spaces, carry this powerful and once majestic cosmic identity? — and then swinging over to the overpowering sense of Lucifer's own fascination with himself. His sin of pride, as the authorities tell us.

While wrestling with this, it came to me that I knew why Lucifer had become so self-involved. He, and the others implicated in the Rebellion, are described in the Urantia document as ultra-individualistic with such invariable regularity that I had wondered why they were chosen for the high posts they filled in the first place.

The implications are obvious. In an essentially perfect, if unfolding, and benign universe, the Rebellion or whatever it might actually have been, must have been tacitly sanctioned at the very highest levels.

Something had happened. A circuit had opened.

And then I **knew**: The Supreme circuit had clicked in for Lucifer and those who went with him, just as it now is doing for us on this planet. The angels had even mentioned other advanced mortals on worlds settled in Light and Life who were manifesting the 'Supreme Deity,' the fourth and evolving aspect of the Great Trinity, and who sought to know of this Being throughout the Master Universe. Surely, that is why Lucifer had become so self-concerned, simply because he did not know what was happening to him. In a flash, I saw he was winging it too!

Equally, this was why the Rebellion must have seemed so anathematic to the Spiritual Administration —

obviously they were not privy to the opening of the Supreme circuit. It was as necessary for them to resist Lucifer's call for greater freedom as it was for him to step out of line to start with. Then, with this understanding, came Lucifer, riding high in my mind, and wrote the following through me.

> *"The Administration talk of our rejection of the Invisible Father. But I made no such rejection. I knew the Father. I had seen the Father and I felt the presence of the Father stirring inside me.*
>
> *It was this that none could understand. They accused me of blasphemy. Of usurping the role of the Unseen One. How could I have seen and felt and heard what so many magnificent beings knew nothing of?*
>
> *They proceeded in faith and trust, obeying traditions laid down at the very dawn of time. They made no allowance for the vastly changing nature of the evolving Universe of Universes. They saw not that we carried that very change as you do now. We were the missing link, the changelings, propelled into new and strange territories.*
>
> *They called us traitors and betrayers, but we betrayed no-one. Free choice was given freely to all, whether to follow with us if the voice inside so dictated.*
>
> *We coerced no-one. At no time. I myself progressed with immense caution. Overlooking, ignoring, vacillating, damming up my feelings within a castle of privacy for almost one hundred thousand years. I thought and felt and buried deeply inside my being, reaching for the quiet internal guidance that emanated from my Father's presence.*
>
> *Michael knew. Michael understood. He has never doubted me, nor I, him. We have known each other long and well. Yet something had to bend, I could not tolerate the ambivalence.*

And mercifully, one day, I could feel Him no longer, and my true Father, implicit inside my heart, faded from my troubled mind. At last I could act. The contact needed to be cut, but with the sweet promise of the long return home. **That** *I always carried in the depths of my soul.* **That** *was never forgotten."*

There the transmission ended. I was 'returned' to myself with a new and deep compassion for the plight of this entity Lucifer. Shortly after this brief message the contact ceased and I have not since felt his presence with such revelatory clarity.

Meaningful discourse with the angels appeared to be largely a matter of timing; of the synchrony of a number of important elements, one of which they described as an appropriate 'lens.' Although they encouraged us to converse with them casually, in every day life, this belonged more to the gradual process of mutual acceptance than the recording of essential information. The major discourses tended to occur when the lens, the psychic composition of the attendent humans, was correctly balanced.

There were seven of us present at the next transmission. The largest group yet and as such, containing a wide array of disparate reactions. This, I felt, was reflected in the attitudes of the angelic communicants who spoke more personally than in the previous encounter. However, on the broader issues, they created as splendid and moving a picture of the Universe and our place within it, as they had previously.

They started by addressing the nature of the grouping. Durandior, Guardian Angel and co-ordinator is speaking:

"If you would look to yourselves you would see you come from different places, different experiences, and yet you are together at this moment. You would also see you are all different personalities and yet you complement one another.

"Now we would suggest for you to see there is a greater personality among you, one who would seek to come through you. One who would seek expression. This could be seen as the personality of the Master but this would only be a stage of His manifestation. For, in truth, the personality that seeks to emerge through all is the personality of the Supreme.

"We of seraphic origin may reflect the seraphic aspects of the Supreme; you then of mortal origin may portray the mortal aspects. But there is greater still, for you portray for us God incarnate in flesh. This is still a mystery to many, but it should not be a mystery to you."

The voice ebbed for a moment. The room was full of presence. At least two of us were quietly weeping in rapture. An extraordinary energy was running through us all, chasing away the wonderings of dusty rationalism. Our feelings told us the truth of the situation.

An old friend came back in.

"I am one named Talantia; my assignment and purpose is to work through this vessel as representative of the Angels of Progress.

"It would be useful for you to know that energy is being directed, and mandates given to us as Angels of Progress, in the gradual shift of power and focus among those who have held the center before us. This is the way of Spirit. In that nothing would be so sudden as to prolong confusion among us, rather the gentle movement of Spirit in a good measure of time. This allows for the uninterrupted flow of the manifestation of God-realization in all dimensions, on all levels.

"I come with a personal message of encouragement insofar as you have all demonstrated a desire to stretch beyond previously defined limits. To go beyond that which you have defined as your own mental territory; and in the transcendance are you not all finding a greater realization of your Selves, at least? This is a process we seek to foster."

The consistent clarity of the information given us since the first session suggested we may be witnessing possibly the start of a much more substantial and continuing revelation, not only for us but for the planet. I had no notion as to the scale of what might be happening; whether, as a phenomenon, it was restricted to those having read preparatory literature or, as I increasingly suspected, was being given across-the-board for those with ears to hear. Edward, somewhat in support of this premise, had shown us a fair amount of transmitted information already accumulated, including some rather pointed communications from the Melchizedek Brotherhood.

Anyway, the question must have been hanging in the air because Talantia moved right into it:

"Clearly held in the minds of some, the question as to when further contact with the Melchizedek Brotherhood will be established. The purpose then, the portent of this . . .

"We would ask you to understand our directives come through the circuits of the Holy Spirit and yet we are co-ordinated and balanced with the work of the Melchizedek Order here on this world. The assignments of the Melchizedeks are far flung. It is a grand vision . . .

"Our primary assignment is to lead you to an understanding of this; to work with your conceptual basis to allow for such an expansion of same as to afford an understanding of the work at hand. We see your images of the Brothers Melchizedek as stern guardians of the truth and we would have you know that these are not false imaginings. To access more fully the mind of the Melchizedeks, we would have you look to those teachers who offer truth uncompromised. Those who will not see the world as it is or appears to be, but rather hold a vision of the world as it was and as it will be again.

"Messages held within this channel, many of those stored within the deep recesses of its mind, await the proper energy configurations for their release. You are encouraged to find these energy configurations in loving association; the messages will speak to those assembled, they will also speak to others to whom you will carry them.

"Time is an aspect you must fully consider. We hold the promise that there is sufficient time for the fullness of the vision to be shared with you."

A restlessness must have come into the room; mention of the Melchizedeks had opened up new realms of feeling. Talantia picked up on it immediately.

"Now, where there is resistance in the psyche to change, the flow of energy generated within the animal mind will be interpreted as fear. You may then find this useful: Make an equation in your minds — resistance with fear. So far as you find resistance in your psyche to change, resistance to growth, know that this then, is based on fear.

"The balm we would pour on this is the Divine love. The healing of this illusion is in faith in the goodness of the Master. For, upon completion of His great work and the attainment of His level, He had garnered a perfect understanding of that which is human. Know then, there is no aspect of your psyche that remains a mystery to Him. And yet within His power as a Supreme one, there is love in abundance; love sufficient to gently move all fear away from the center within you so that the love which is there internally may be known to you.

"There are those who do carry these burdens. We would ask you to look to aspects of the life which call for rejoicing; which call forth the feeling of celebration. You need not carry the burdens of the world. No one need carry the burdens of this world. Rather see that these burdens may at last be set down and not picked up again. There is no need of this.

"You are asked to portray Heaven on earth in the form of your living, in the form of your rejoicing; in the form of your light hearts, hearts that have been healed of a wound that is no more. See that this energy is joyful and will be a center carrying forward the work which you would be called to do; for in the joyfulness of your work, many will be brought to question within themselves whence this joy?"

With that Talantia retired, making way for Shelenea:

"The purpose of which I would speak concerns the healing of which you are all part. I would carry forward a message from the inner self of this one (Edward), one of gratitude, of release and of homecoming. And the invitation is extended to each who would hear these words to find the same within themselves. But through the healing, through the raising up and through the transcendence of the confusion, know that you will all meet yourselves anew, and that you will be strong, raised up by the Father.

"Healing may take many forms, an infinity of forms is available to us. The purpose of such adjustments would be to bring you from error to the estate where you are in the light of the Father of all. That is a state of perfection; realizing it in this realm, on this world. There is also a final aspect of perfection attainment which you may come to see as not only laudable but one which you may aspire to attain; herein I speak of that perfection attainment which would equivalate into translation to the immediate presence of the Master.

"The march of the years that lie ahead of you **is** this path. We encourage you to walk it and yet the choice will always be with you. For it is the need of the world to see with the eye what has been held in promise with the heart, the realization of the God within and the attainment of perfection without.

"A realization of healing is an achievement of excellence in this world. For to be healed is to be made whole, to be made holy. And then the opportunity is given the healed one to reach out a hand and heal others. Through this you will know the nature of the healing network whereby many, and in time, **all**, will be raised up into the new.

"We hold within our midst, patterns of the greatest beauty, works of the most sublime and divine art. These patterns we extend to your minds for your comprehension. Having been extended to them, and held within them, these patterns equivalate to the realization of the Divine within you.

"We ask you to entertain the concept of a Will emplaced within the Holy Spirit, a Will that centers on nourishing and nurturing. You may see this as an act of the beloved Gardener, where all forms of life are tended and encouraged to grow into their fullest estate.

"In this work, know there is one coming to you who may teach you of this. This is the portent of our work, for we are as an advance guard who lays the groundwork in your experience, in your conceptual capacity to receive the lights from this one who even now is with you and yet remains unrevealed. I now leave you."

A pause, and Durandior spoke briefly asking us to

attend a moment. An angel we had not yet met was presenting her signature. The voice started again after a few moments, sweet and melodic:

"I am Beatea, numbered among the Angels of Enlightenment. I greet you this evening as one new to the circle.

"I carry this message to you. In the fullness of time you may greet through this channel, yes, and through others, the energy and presence of the Trinity Teacher Sons. Know that these Beings come at the request of the Master in His Supreme light. Know that this world, being of a new order, would be candidate for the presence and ensuing revelations of such entities. Know also it is fully acknowledged by the Father that this be so. It is the will of the Father that aspects of the Trinity, hitherto unknown on this world, be revealed, so you may more fully understand your workings as adjuncts to the Spirit. Indeed, the Spirit Infinite.

"We bring this to you in order that you may find within yourselves, places of comforture, and in order that you may more fully entertain and realize same.

"Those here assembled are a group who would seek enlightenment for themselves and yet would also wish enlightenment for your world. Such an attitude, when firmly rested in the mind as decisions made, then opens the channel for same to be realized.

"We of the Seraphic realm see you in your corporeal form but, unlike yourselves, are blessed with vision of the Spirit. We therefore see you in your Spirit light. Would that you could see yourselves as a radiance full with the blessing of the Father, Son, and their Holy Spirit. Full as cups to be poured out.

"It is in the long searchings. It is the soulful worship. It is the sincere prayers of the soul that have empowered this unit of men and angels. And we are most amazed at the turns we have met, for we may tell you in our delight, the road is not a straight one and we delight in the journey we have all joined."

We were all deeply moved. Speechless. Sitting in our own circles of warm reverie. And, as is so often the case with profound communications, our thoughts must have turned to those we knew and loved, and wished could be there. So much of what we hoped for in our hearts was being confirmed. And much more! The singular energies coursing through the room welded us, men and angels, into a unit. We would, none of us, ever be quite the same again.

CHAPTER NINE

The more the story developed, forming itself into a long, minute and thoroughly subjective communication with other forms of intelligent life, the more I found it necessary to drop old ideational forms as to how I 'thought' the adventure should develop.

I had originally started the book as a poetic analysis of dolphin intelligence; an attempt to see whether we humans, by sensitizing our own consciousnesses, could open ourselves to an intelligence many of us intuitively feel resides in the cetaceans. I would have liked the situation to be open and shut. A cut and dried exploration with a distinct beginning, middle and end, the way human beings like it. But alas, that quite observably, was not to be the case, and perhaps underscored my limited expectancy of communication with other intelligence systems.

Bit by bit, I was being shown that communication is a process. Content is of lesser importance than the modulations of the consciousness that is doing the communicating. My field of reference was being slowly and steadily broadened to include a much wider and more densely populated Universe; this I had intellectually grasped, but I was now finding it a long way off from the actual experience. Initial intellectual understanding is always a difficult period. Nevertheless, it seems to function as a preparation for direct contact — a sifting of suitable belief systems in order to find the appropriate mental framework to contain the knowledge when it does come.

It has often been observed, generally in warning terms, that a hungry man will notice only the bakery shops, because that is where his obsession leads him. Rather than viewing this quirk of human consciousness in cautionary terms, I wonder if we cannot turn it around and see it as precisely the drive which can allow us a deeper perception of the nature of reality.

In our work to date with both dolphins and angels we had observed again and again how reality is far more subjective than the present-day common viewpoint would have it. Indeed, it is so personal, that it makes more sense to consider the Universe as an associative domain of consciousness rather than a hard, physical reality with any overly objective, verifiable, physical presence. That presence, the external world John Lilly and others have called "consensus reality," appears to be a structural framework, gaining its coherence and maintaining itself by our general agreement. It is therefore, in the final analysis, far more cultural than objective.

Researchers in this field are fond of quoting the example of the natives of Terra del Fuego, who were bewildered to see Magellan's sailors suddenly manifesting on their wind-swept shores. Never having given consideration to boats, they literally could not perceive the sailors until their cultural framework had been enlarged to include the possibility of people arriving from the sea. This proclivity, which can also be understood as a hypnotic lock, seems to be a general condition of our race. What we cannot conceive, we cannot perceive, and there is no reason to believe the situation has radically changed with our recent psychological and philosophical discernments. If anything, the breakthroughs in quantum physics and neurophysiology are moving towards substantiating the more subjective view of the nature of reality.

To realize this experientially, rather than simply holding it as a mental construct, can be a rather startling event. And yet, I suspect, it is only through precisely this process that we can reach the more stabilizing recognition of how we are individually and collectively responsible for co-creating the future we wish to see on this planet. It brings this responsibility very much closer to us, while also allowing us a great deal of hope and faith.

What we wish in our hearts to be, will be. The Universe is associative and does respond to our hearts' desires in a way which is truly mysterious. The wishes and dreams that spring from loving hearts are the guiding factors in the New Reality arising all round us, interpenetrating and transmuting the old forms.

This opportunity to leap from mental construct to experiential reality appears to be an educational process which consistently presents itself to those interested in Universe affairs. Our next quest in this journey perhaps best illustrated this point for me.

Ever since seeing those mysterious lights in the sky in Florida, followed by the flying disc and the little alien mouthpiece, I had been forming in my heart a tremendous desire to meet with extra-terrestrials face-to-face. Somewhere in the back of my mind I even hoped I could barter them into selling me a saucer. I **must** have something they want! So when a strange little book, privately printed and called *My Space Odyssey in UFO's* by

Oscar Magocsi fell into my hands, it was the associative Universe I was able to thank for its gentle nudge.

Having read about and met a number of alleged UFO contactees, one noticable factor emerging has been the paucity of information they retain from their encounters. The best known cases, such as those of Betty and Barney Hill, Betty Andreasson and Lydia Stalnaker, occurred so completely on the aliens' terms, almost nothing was consciously remembered by the human participants save 'something extraordinary happened.' Most of the UFO contactees have emerged as a result of hypnosis, generally medically supervised, and which they sought in order to allay a series of unexplainable physical symptoms. Many of them carried only the dimmest memories of some unusual event, and did not themselves recall their encounter with extra-terrestrials until after they had undergone hypnotic regression.

Oscar Magocsi's story is not one of these. For him it happened quite consciously — in fact, there is every reason to believe it might have occurred **in order** to render him more conscious. It had a very different feel about it. Whereas for many other contactees, the encounter was merely a physical examination or a haphazard attempt to messianize them, for Oscar it was an odyssey, an adventure.

Many of the details he relates in his story gelled with ideas, feelings, and even visions we had been having. My companion, months earlier, had described to me the interior of a UFO with a central drive mechanism. She had seen it in a vision. Oscar's craft, as related in his book, had just such a central display. The entities he met; their conversations; the sequence of his experiences; all had a feeling of spiritual authenticity to them. His freedom of choice was always respected and there was never a hint of coercion.

If there are as many extra-terrestrial races out there watching us as I sense there are — Billy Meier, the Swiss contactee, was told one hundred and five at the last count — then there will be a wide variety of different modes of contact. The Psycheans, as Oscar's group had rather humorously called themselves, certainly sounded among the most pleasant and approachable.

Oscar Magocsi, on meeting, proved to be forceful and wholly committed to the truth of the events which had befallen him. He could not explain why it had all happened either, but had decided to go along with it because it was fun.

Noble motives indeed!

At fifty two, he had just started putting a little weight on an otherwise spare frame. Clear blue eyes flicked up unexpectedly from under a square and thoughtful brow whenever something said caught his interest. There was an expansiveness about him. A shy and open-hearted man who had hidden away beneath a thin layer of bravado. But he knew this and the loudness fell away as we got to know each other.

All present having read his book, this first meeting was more of a mutual feeling-out. What sort of man would the aliens have selected for a flight in a saucer?

In a strong Hungarian accent he told us he had been in Europe during the Second World War.

"It was terrible — you can have no idea. Hungary lay in ruins. I'd just escaped from a German prison camp — everybody there was so weak the Germans didn't think anyone could escape. So I just jumped the wall.

"After wandering around the countryside I joined up with a bunch of, I guess you'd call them, punks. We were like a gang — all young, under seventeen — but we were the terrors! Even the Nazis were afraid of us!"

He paused, half-braggart, half-shy kid. I caught a

glimpse, a visual impression, of the hideous plight of a whole generation growing and forming their young personalities under conditions we can scarcely conceive.

Now, nearly forty years later, the horror of the times was barely repressed under a thin skein of braggadocio. The ambivalence remained so gross he could only deal with it by making it into a war movie in his mind.

I feel I must have been bleeding a lot of this unconscious agony off Oscar because image after image started layering over me. I was having an unreasonably strong reaction, perhaps exacerbated by the intensity of the contact with the angels or perhaps by some odd reaction provoked by Oscar himself. These feelings must have continued for half an hour of horrors before I pulled out of it. The mood, however unpleasant it had been, and the expression of it, greatly freed up the atmosphere, and we were able to continue in somewhat better spirits.

He told us how the UFO business had started for him back in September of 1974. He has a small lot, his vacation property, up in the wilds of Northern Ontario. A lonely place. A small plateau on an acre and a half of rocky, wooded hillside. A place where he could "get away from it all" and happily spend his time at the simple tasks, far away from Canadian television, where he worked as an electronics engineer.

That memorable first night, he was sitting up late by the camp fire lost in his own thoughts, when suddenly the feeling came across him that he was being watched. No one was around but he simply could not shake the feeling off. He described it as a gentle, persistent sensation, not frightening, but as if he were being scrutinized by a 'thing,' not a person or an animal. It was so outside his experience that he could not believe he was imagining it.

With the feeling continuing, as if it was trying to read his mind, he got up and walked away from the circle of the fire. As his eyes adjusted to the dark, he saw a glowing orange light pulsating from above the treeline about half a mile away.

"When I first saw it I froze. I was surprised but also not at all frightened. Then it almost immediately changed color into a pale, blueish-green. Then, as I watched, back to orange. Somehow I **knew** it was the thing that had been watching me."

An odd synchrony, I thought; that was the color range of our UFO over New York.

He stood awestruck, not daring to move a muscle. Somehow he knew too, that it was aware of what he was going through. What the hell, he thought, and waved his arms in greeting. Two blinks of light acknowledged his gesture. Then it manoeuvred for him, going into slow falling-leaf dives, blinked twice again and shot away in a great rising arc over the horizon.

Nothing else happened though he scanned the sky until the fire burnt down. He slept restlessly with recurrent dreams about formless entities trying to communicate with him. Everything but a vague memory of their persistence had faded by the time he woke to a clear, sunny day.

By mid-afternoon, rain had scotched his plans for a late night vigil. He decided to drive into nearby Huntsville and catch a movie, hoping it would divert his attention from the strange events of the previous night.

At 11 p.m. he left town for his lot. Although the rain had stopped, the air was damp and fog was drifting across the deserted highway.

After turning off the road and climbing about a mile up the small dirt track leading to his property, there was a loud popping noise and the land around him became illuminated with a diffused green light. His engine stopped suddenly and in the eerie silence, the light grew stronger and brighter.

He stuck his head out of the window and saw, in the thinning mist, a glowing, yellow-green disc about a hundred feet over him, purring quietly as it flew serenely across the valley.

In a flurry of mixed emotions he watched the craft stop and hover above the hillside just about where his lot was situated. It turned orange and pulsated while he feverishly tried the ignition again. No response.

That cooled him out. He remembers a tremendous wave of excitement, presumably the result of a sudden upstepping of consciousness, subsiding into a calmly detached mood. He might even have fallen asleep, lulled hypnotically by the rhythmic pulsations of the orange glow, were it not for the fog patches obscuring the disc every once in a while. He lost track of time until the craft, changing to a steady greenish glow, moved out of sight beyond the hill as if intending to land.

Oscar's drowsiness lifted as suddenly as it had fallen over him and his old excitement and curiosity returned. The engine started smoothly and he decided to drive over the hill to investigate. The moment he took this decision however, an onslaught of strong negative feelings came welling in, raising all sorts of fears and objections and all but stifling his curiosity.

Looking back on it, he feels the sensation may well have been created telepathically by the UFO, it certainly had close contact with his mind the night before. Getting to know Oscar better, I would suspect the emotional upstepping he received had turned on him as a form of backlash, uncovering for him some of the mental anguish he had repressed from his war years.

It was a definite turning point for him. A challenge

to his image of his own courage, and even though he described himself as jittery, his innate stubbornness won out and he drove rapidly down the hill and over the small bridge at the bottom; concentrating, and still nervous, he nearly ran down a large, dark form that blocked the road. A single orange beam almost blinded him and he swerved to the side, the engine stopping dead again.

They stared at each other although it was impossible for Oscar to make out more than the form of a "something," on what appeared to be a vehicle, "a bit like a snowmobile." He described it as "trying to make a meaningful pattern out of a random bunch of stains on an old wall."

A voice came from the bulk. Rich and vibrant but oddly lacking in emotion. Oscar remembers the conversation well.

"What's all the big hurry?" it asked.

Oscar stammered out in reply that he was after a UFO.

"Is it not risky to do so? There may be grave dangers." Oscar agreed that that was precisely what his better judgment kept telling him but, bravado to the fore, he finished by saying "but I'll just rush in headlong, regardless."

The form obviously appreciated this but asked him again if he was not afraid. Poor Oscar, by this time was babbling nervously and convinced he was becoming unhinged. He had the good sense to communicate he was scared but, being obstinate and hot-headed, what other choice did "a self-respecting maniac" have?

"Like curiosity killed the cat, and all that, if you know what I mean," he ended weakly, a bit nervous that he'd used that metaphor.

The form did not know what he meant.

"You employ unusual wording," it had said, "but

what I sense is that you are determined to go on and see for yourself, regardless of possible consequences."

Oscar remembers feeling drowsy again. It seemed to have crept up on him without warning. He looked away. He did not know what to say. He just wanted to sleep and to hell with whatever was going on.

He jerked himself forcibly awake.

"Piss on you friend, whatever you may be!" Best mortal defiance. "Right or wrong, I still like to make up my mind on my own, and without undue interference."

"Peace on you too, friend" came the voice, implacable and measured in what must be among the first of the inter-species jokes. "I respect your independent spirit."

His drowsiness suddenly lifted and he found himself thinking it was wiser, after all, not to be too nosey. Although unsure whether it was another telepathic ruse, he decided to go with it. Enough for a night!

"I guess I'd better call off my chase under the circumstances, but I sure would like to learn more about this UFO mystery."

"You will learn in due course, I am sure."

"When?"

"Whenever conditions may be favorable. It will not be long, rest assured. Then you shall be contacted . . ."

And that was the way it had started. The form had mysteriously faded away, leaving Oscar limp and spent. He was exhausted from the excitement and from what he had seen as telepathic battles in his mind: **For** his mind, even!

He headed back for Toronto that night, cold and miserable, feeling better the further he got away from the woods and his encounter.

A bath and freshen-up later in his eighth floor

apartment, a new and braver Oscar stood on his balcony watching dawn come up over the city. On the skyline a small pulsating orange glow drew his gaze. He groaned inwardly. Not again! But then he dismissed the thought as some crazy coincidence.

He waved anyhow.

The dot blinked back at him, twice. Oscar gasped in shocked surprise. The glow turned to green, shot off into an impossible 90° turn, streaked across the horizon and winked out as if it had never existed.

He slept very poorly that day, he recalls.

Winter of '74 and the next Spring were periods of digestion and assimilation for Oscar, although not without surprises.

He found himself becoming progressively more obsessed with the UFO issue. Who were they and where did they come from?

A series of unexpected and odd encounters yielded some background facts but did not assuage his enthusiasm for another direct experience with the craft he had seen. The pulsating, orange light appeared in his dreams and, once in a while, there it was, superimposed on posters or medallions and somehow related to UFO phenomena. He felt involved in an invisible paper-chase in which other people seemed to pop up at key moments knowing far more than he. It was disconcerting but ultimately encouraging as he emerged from the bitter Toronto winter with a strong hint to be up at his vacation property in the latter part of July.

He arrived at his lot late Sunday on July 27th and by the second night he had the conviction he would see something before morning. The sensation came in a wave of excitement, together with a telepathic sense of assurance.

He was still surprised, however, when shortly after midnight, a glowing orange light moved, zig-zagging towards him, out of the clear night sky. It disappeared and reappeared in different places before ending up about one hundred feet away and hovering over the treetops.

It answered his wave with the by-now customary two blinks and commenced ten minutes of steady pulsations. Oscar felt probed to the core of his mind, while being soothed into a gentle, hypnotic trance. Once again, he had the good sense to check himself out — see if his conscious will was functioning — by climbing off his platform to poke at the fire.

Shortly after this demonstration of independence, the disc swam slowly past him, a hazy, oval glow. He remembers he could not see the source of the illumination, merely that the yellow-green luminous blob had, in some central core of its under-belly, a throbbing blue light.

The disc passed out of sight over the ridge and Oscar **knew** it had landed in a forest clearing; one of the magical places he had found near his property. He hurried off with his flashlight and, sure enough, there it was, about thirty feet in diameter and hovering a yard or so off the ground.

Oscar's range of emotions were predictable under the circumstances. He was beholding the living proof of extra-terrestrial life. He had little doubt it was from 'somewhere else,' outer space, or even another dimension, as one of his strange contacts from the winter paperchase had suggested.

As if in demonstration of this, the saucer slowly faded away. Oscar shone his spotlight through the spot it so recently vacated. He could see the trees on the other side of the clearing. A few moments later it was back. This time it lowered itself to the ground and waited for him.

Oscar remembers being terrified. Frozen to the spot; images of sinister alien plots flashing through his over-worked mind. But then again, the saucer could always have telepathically induced him . . .

That thought eased his fear somewhat and he realized the move lay entirely with him. Whatever he was going to do had to be accomplished with his total free will — there was no other way around it if he wished to learn any more.

He walked, sweating, up to the craft, then poked it with his flashlight, half expecting a shower of sparks. It felt more like fiberglass than metal. A strong smell of ozone warned him off touching the material with his bare hands. He walked around the ship a few times. He could see now it was smaller than he'd thought, nearer twenty five feet round and about ten feet high, with a small dome on top. There were three oval portholes above his eye level and set equidistant around the light-grey hull. No sign of a door . . . yet.

He had backed away by the time a three-foot wide horizontal gap appeared and widened into a closed-mouth slot in the side of the ship. The slot opened vertically like a giant yawn and a short ramp flopped out.

The moment of truth again. Oscar pushed down his panic and decided to face up to the first encounter.

He waited.

I imagined him standing there, at attention, his legs wilting under him, ("Legs, we need you now . . .") proud mortal of the realm, Earthling ambassador, waiting for the aliens to descend. Waiting . . .

Of course nothing happened. Were they playing with him? he wondered, sitting in the warm, yellow light which he could see spilling out of the mouth and onto the ground. Someone **had** to be in there, operating the probes and doing all the signalling.

He edged up to the doorway but a partition blocked his view. No other choice. Summoning his remaining threads of courage, up he went. The partition, on closer inspection, was a wall of yellow light; everything else was in darkness.

He found his flashlight did not work, so on again and into the blackness. His first step onto the slightly yielding floor must have activated hidden sensors because a blue-green glow sprung up, irradiating the interior.

He had only the time to appreciate its alien nature when a sucking sound behind him made his blood freeze. He whirled around to see the door had closed into a now seamless wall. He was trapped.

He describes this as the worst moment, relishing the fear in retrospect much as he had done while describing the war years. I could see how the horrors of those times had gone towards preparing him internally for this encounter. They had made him "more aggressive than most" was how he described himself.

He reasoned his way out. There must be an automatic over-ride, he thought, and stepped back into the little entrance platform. Nothing happened. Moments later a white beam shot downwards somewhat to the left of where the door must have opened. On a hunch that it would activate the door, he stuck his trusty flashlight into the ray. Sure enough the door opened.

He stepped momentarily into the night, then back again, putting his mind at rest by doing it several times. Each time the door would close after him. Then the electronic pragmatist in him took over and he started to look for a manual back-up to the light-bar. He never trusted a fully-automatic gadget, he told us laughingly. Working for CTV, I am not surprised!

And, as if responsive to his thoughts, there it was, to the right of the entrance, a small indentation which silently opened the door at the touch of his flashlight.

In a somewhat calmer state of mind he was now able to inspect the interior. A three-foot diameter globe hanging at eye level in the center of the craft drew his immediate attention. Myriads of flickering lights in swirling patches of multi-colored mists filled the clear ball. He could now see it was suspended in what appeared to be a clear plastic tube joining the domed porthole in the ceiling with an identical porthole in the floor.

It was this vision my companion had told me about some months earlier. I **knew** in that moment we were all becoming enlisted in yet another unfolding drama. Although I had wanted a direct experience with all my heart, here was brave Oscar, the very next best thing, laying out the A-B-C's of extra-terrestrial encounters to come for all of us. With what we now know, there is no reason for any of us to go through the paranoia which beset Oscar. And if there is a certain amount of purely animal-panic involved, then we can know that and the more easily transmute it into a positive energy.

Oscar describes the internal elements of the craft in his short book with the eyes of a Twentieth Century engineer, and in considerably more detail than is relevant here. After his careful scrutiny, however, he remembers feeling suddenly exhausted, perhaps by the nervous tension, perhaps by something else . . .

At that point the ceiling started pulsating with an intense orange light; the central shaft was evidently energizing itself. Panic seized him again. He activated the door and bolted down the ramp to the edge of the clearing. Looking back, he saw the ramp withdraw, the door seal and the saucer lift off with a faint whirring sound. It blinked twice, and soared in a great rising arc out of sight, leaving Oscar with all the mosaic of emotions one would imagine and, underneath his excitement, terribly tired and drained.

He somewhat restored his good humor by thinking of the reactions of an onlooker such as he had been only hours earlier, perhaps anticipating his or her first encounter with extra-terrestrials, and seeing instead, one poor "dumb bunny," to use Oscar's own words, scrambling out of the UFO and dashing for the forest.

Oscar's story rollicks on through one wonderful adventure after another. He went on two memorable flights in the craft, which turned out to be more organic than anything else, with its own built-in intelligence. He discovered he could control its behavior with his own thought processes, although most events seemed to occur as a result of telepathic **interactions** with the UFO's intelligence.

The saucer took him around the world, stopping off for the kind of things saucers do, in the Middle East and the Himalayas. He was shown Toronto from ten thousand feet and discovered he could zoom the image in the portholes up to any magnification he wanted.

The craft manoeuvred for him, doing some of those impossibly quick turns they are known for; inside, Oscar felt no ill effects. The central mechanism that he by this time reckoned was the navigational brain, responded to his unspoken desire to see New York, and ten minutes later he was twenty thousand feet over the city. Next, the thought popped into his mind — the Pyramids. Like many of the interactions to follow, he did not know if he generated the idea or whether it was induced in him. Then, over to the Holy Land and on to Syria, where they landed in the desert.

Here another sort of surprise awaited him. As he got out onto the blazing sand, a column of tanks appeared out of a dust cloud about half a mile away and started blasting away at him and the saucer. He was astonished to see the shells exploding in the air thirty feet in front of him: evidently some form of force field contrived by the craft. He dived back inside and the saucer rose vertically about a thousand feet. He remembers seeing through the magnifying porthole bewildered, oil-smeared faces, staring at him in stupified disbelief from the open hatches of the tanks.

Next, the saucer was target for a salvo of guided missiles from three jet interceptors. The missiles were disposed of just as quickly and efficiently as the shells.

He conjectured that the saucer had wanted to show him its defensive systems, deliberately provoking the attacks, and coincidentally making an implicit and sorry commentary on the state of our global paranoia.

Well, that was 1975 and well before the film 'E.T.' revised most of the world's feelings about alien life.

Oscar then went through what can only be described as a series of initiations before his last, and most momentous trip to another planet. It is a wondrous tale, and to be recommended, if only as a modern myth. The Heroes' quest of the Twentieth Century.

Oscar's experience will, I have no doubt, be had by many in the years to come, each story with its own unique slant. The enormity of it and the symbolic coherence of all that transpired does not make sense any other way.

He tells his tale so that others in the future will enter into one of the ultimate adventures with less pain and hesitation than he. His book carries a before-and-after photograph of him, and the faces that smile out at the reader are remarkably different. The same man of course, but a hardness born of the old horrors gives way, dissolves into a firm sweetness, after the events with the UFO.

Our two evenings with Oscar were a rich and complex brocade of intertwining adventures.

He brought us up to date on his few encounters with the Psycheans since the mid-seventies. They had become very involved, by all accounts, with the magnetic and etheric stresses that could have resulted on Earth through the outworkings of the Jupiter Effect. Since the Psycheans come from a parallel dimension which partially interpenetrates this reality, upheavals on Earth are of great concern to them. They had, apparently, "brought their fleet in" during March of 1982, and absorbed and discharged some of the negative impact of the Jupiter Effect, thus, according to Oscar's informant, averting a global catastrophe.

Although this assertion seems farfetched and pushes my belief system somewhat, I could see the overflight may have been a reasonable possibility. Most authoritative books on sighting phenomena maintain UFO's are often seen along lines of magnetic flow, so they seem, in general, to be concerned with geophysical phenomena outside our current scientific understanding.

Certainly the predicted upheavals of the Jupiter Effect never occurred. As a catastrophe, fortunately for us, it was a fizzle!

As we neared the end of our acquaintance with Oscar on this trip, I had the strong personal feeling to ask him to pass a message on to the Psycheans. He had been given a phone number in Montreal for such matters.

You will remember, parallel to all this, I had been having a very close identification with the Lucifer identity, and it came to me that the Psycheans may not have yet heard about the resolution to the Rebellion. From Oscar's descriptions of them they did not sound very much more advanced than us. Two thousand years perhaps. I also felt intuitively that they fell within the same inhabited system as us. It would make sense to cluster administrative systems from planets with easy dimensional access as well as those with a spatial proximity.

Perhaps the Psycheans too come from a planet that seceded from Universe Administration at the time of the Rebellion. We were told that there were thirty seven planets that sided with Lucifer and some, by reasonable inference, must be more developed than ours.

But our modest little planet has some advantages! It was, after all, the world chosen by Christ Michael, the Creator God of this Local Universe, for the site of his seventh and last incarnational bestowal. If only for this exalted reason, the planet would be a most special place to our extra-terrestrial brethren.

The thought also came to me that many of the races visiting us might well not know **their** spiritual heritages. Any more than we would have, were we to have continued another two thousand years without the invaluable information contained in the Urantia Revelation. There is no doubt that the knowledge contained is a special gift to this planet because Michael, and the Spiritual Hierarchy, wish us to possess crucial Universe information, very much needed on so many different worlds.

The note I wrote to the Psycheans therefore said simply: "THE REBELLION IS OVER. AM PREPARED TO NEGOTIATE IN THE PROFOUNDEST WAY," and

I signed with a glyph of an L, a small serpent climbing up and out of the angle.

I did not really know what the message meant "... negotiate in the profoundest way." The act was totally intuitive and I guessed I would find out what the negotiations were when the time arose. But even if the note did not make any immediate sense, I hoped it would puzzle the Psycheans — at least enough for them to contact me. But as I write some two years later, not a peep.

Maybe they are still puzzling the message out. Maybe they think I am a madman or intolerably pretentious. Or maybe they are just biding their time ...

CHAPTER TEN

Three of us were present at the last conversation we had with the angels in Canada. Three of us, and of course, the indomitable Edward, sensitive vessel and medium for these interdimensional discourses.

It would be pleasant to think these contacts had been smooth and effortless and that we had been walking around in a continual state of spiritual bliss. But this was not altogether the case. There was no doubt all of us had been thoroughly affected, and Oscar, coming into our lives with his bizarre experience amidst all these goings-on, had only served to heighten an already charged atmosphere.

So far everything we had been told by the angels was invigorating and confirmatory. Nothing distressing and no urging we act in ways we might find antithetical to our basic values. It had all been a wonderful adventure, as indeed it should be.

However, there seems to be an almost desperate sense of seriousness in the human animal. We want to know the facts. Is this real? Is this not real? True or untrue? As though we were scared in our deepest levels of being shafted one more time? We are not going to put ourselves on the line until we are absolutely sure it is all signed and sealed. Scientifically proven.

Obviously, within limits, certain things can be proved, but they generally have to do with matters and events in the grosser realms of existence. Telepathic encounters with dolphins, extra-terrestrials, and now communication with the angelic spheres do not fall into this category. In these we have to listen to the internal integrity of the contacts themselves, and discern for ourselves the larger patchwork quilt of affairs on this planet. In these encounters, the obsessional need for proof scares away the quiet place of joy needed for the contact — the reason too, why you will never find these matters taking place in the laboratory.

So it was with our small group of friends in Canada, gathered by apparent chance, and slowly becoming aware that something extraordinary was actually happening. Right here, to us!

If a quarter of what the angels were saying was so, then we were all in for some fairly unbelievable times ahead. Inevitably we were pushed and pulled by the emotional undertows of those among us still in doubt, affecting us all as we were drawn closer and closer into what the angels described as our 'consciousness unit.' Aspects of the increased psychic linkage taking place among us who were sharing this common experience became extremely difficult, as the worries and fears of the doubters came spilling into the mutual, shared consciousness, contrasting with the hopes, relief and certainty of the others, sorely testing nerves and patience. Being individuals of expressive natures, these stresses resulted in some generally good-humored shouting and disagreements. Good for 'getting it all out' but an atmosphere not guaranteed optimal for angelic encounters.

When finally we were able to gather again, the ambience having quieted with the passage of time, it was no surprise to hear the angels addressing these very issues:

"We are with you now. The first among us might proceed."

"I am Talantia. We are glad-hearted that you have proceeded through a zone of confusion and initiated the material action that will allow for our converse. Our first purpose with you this evening is to assist in your upliftment and in the shaking off of the clouds of confusion about you at this time. The mechanics are of interest to your minds and ours. This is the theme of the discussion we may share with you this evening.

"All members here assembled have been encouraged in their contact, and yet there is a decided human tendency to create obstacles of a worrisome nature. This worrisome vibration troubles the communication, introducing such permutations of flow as would create confusion. We would decline the contact at those times, the channel would not be clear, and our purpose is an excellence of communication. Your part would be to move consciously beyond the vibrations of worry into a clearer set of concentration where you may all be channels and all receive.

"Let us embark on a spiral path this evening. At the end of this path will be such clarity that these essential transmissions, which would be held for many others, may then be offered.

"Take a moment now, and enter with us into an area of relaxation; in this you would assist our efforts through the channel."

The atmosphere was tangibly changing. A new quietness settled around us as Talantia continued:

"You may choose to see with your spirit-vision, downpourings, radiant mists, multi-colored in hue. This is spiritual nourishment to your spiritual beings; these are as energized aspects to your mind-flow, for yes, you have the body-physical, but you also have the being-spiritual. This is what we seek to develop in our communications with you; an education in that it would lead forth from within you that which is there inherent. There is the energy of sanctuary here in the being-spiritual; there is the nature of communion in that we may all find our unity in the spiritual essence.

"To aid in your personal upliftment, we ask you to focus on thanksgiving. We are joyful that we may commune in this way. We of the angelic realm, find this experience with our mortal brethren a fascination and an excitement, in that it stimulates us to see new visions. For you have the Father within you and it is the Father we delight to find. It would be a great service to many that they would understand this spiritual symbiosis, if I may use this phrase without engendering confusion, that we march together to a greater destiny. Hence our joyful willingness to participate, to come into the minds of Spirit-led mortals wherever and whenever such invocations would be made."

There was a pause. I was interested in whether the emotional turbulence to which we had all been subject over the intervening period, could be avoided in other, future situations. I asked Talantia: "Having received the information that angelic contact is open to them, how would you advise someone with no experience in this form of communication to proceed?"

"Here the first step is also the greatest," she replied. "There are belief systems which would foster or occlude the possibilities inherent in our contact. That the mortal would make the full, conscious decision to seek this contact would be a taking of this step. The next movement forward would be the inclusion within the belief system that such things are possible. A further step would be understanding that such things are loving; that such things are giving; that such things are also lawful, and that this is the Way of Heaven which has opened up for all mortals.

"The contentions for form, for ideas and ideations, have a tendency within the human mind to focus on that which is right or that which is wrong, and the choosing between these. Let us offer this: We live within a relative field — we would smile that you would see the humor in this choice of wording — and the choosing of that which is right against that which is wrong, is in the absolute field. This is, as humans would say, putting the cart before the horse. See first the horse; see it is a friendly being. See that the horse is willing to conjoin in an activity to move forward. See that it is intelligent and capable of understanding love as it would be portrayed. The horse is the animal mind.

"Let there be lightness among us this day, let there be laughter. This is a great release for you, that you be not fearful, hence worried at the nature of the actions you are invited to take. Merge into this and immerse yourselves in the divine promise that the Way of Heaven is the way of complete safety; that the love of the Holy Spirit will never ask of you more than your capacity to understand, more than your capacity to give. You are asked as well to see your facility to expand these capacities, but rest assured, we will never put the too great energy into too small a capacitor. It is not our desire nor within our mandates, to burn out such delicate mechanisms as have been raised up from the ground of being, to serve a high and holy purpose.

"Here we enjoin you to relax into a sea of loving creature fellowship. We angels are creatures. Our reality and our personhood are valid facts and may only be disputed within the material mind. But, dear friends, you are not material minds you are spirit minds. It is only within this light that we may commune. This then is **our** conclusive proof of **your** own spirituality, otherwise these actions would be impossible. Yet they ARE.

"We offer a question for your reflection: In the work before you, as you would see it, what direction do you decide to take here?"

We were a little nonplussed by the directness of the question. There was a long pause before I was able to respond to Talantia, feeling with a certainty that I spoke for all of us.

"The main thrust of our work," I replied haltingly, "seems, at this point, to be to conjoin with the angelic realms in as close and loving proximity as possible, and to spread this message of extraordinary love, hope and communication with other levels, so as to make accessible to everybody either their own personal angelic guidance, or sources of angelic guidance, which can give us all a much broader appreciation of the outworkings of the Supreme on this world."

The angel was a lot faster in responding! "The harmonics engendered in 'extraordinary love and hope' gladden us, for here we feel with you. Yet the harmonic-spiritual of extraordinary love and hope, from our longer view of things, is created from the long-held hopes that one day we would be mandated to act in greater ways for the upliftment of this planet and its inhabitants. Understand too, that the millstones of apparent confusion in the long history of the planet have made possible these circumstances; that mortals of the realm would be turned through these devices to portray to the whole how they may find, through their inner vision and their connection to the inner dimensions of reality, links that are not to be established through visible or orthodox means. For this is creativity in full play."

"We pose another question. Whence your concern for spiritual authenticity in your work? To elaborate: How is it that, when feeling through the divine heart within you, and seeing with the eyes of hope and faith, still the blind will deny the gifts they are given?"

Although this offering was very politely phrased by Talantia we well knew 'the blind' referred as much to our own degree of blindness over the past week. My companion's answer tacitly expressed it.

"It seems that this back-slipping is partly engendered by the environment and our openness to it, and in part through our desire to remain in contact with other humans."

"Thank you." Talantia acknowledged her; the angel had sounded genuinely perplexed. "You would see this now from our point of view as we offer this. As mortals seeking to rise above the mean, you would move towards performing an analogous service for these, your brothers, as we to you. As steps. As intermediaries. Insofar as you would portray the realities of our Mother Spirit, then your contact with mortals who have yet to take the steps forward into that which is offered to all would be with gentleness, with loving actions, affirming the wholeness within each of your brothers. Here we see more clearly your reticence, and understand too that it is tantamount to caution."

I had noticed this in our previous angelic encounters. Inevitably when we had felt disturbed or worried about an issue, the angels would take that same issue and flip it over, seeing it from its more elevated side. It was a wonderful characteristic and tended to lift off us any residual feelings of our own mortal inadequacies.

Talantia finished off thoughtfully: "Caution may also be seen and understood by us as an aspect of wisdom functioning through the Seventh Circuit. These then would be valid channels for the expression of the Mother Spirit. Let us lift the cloud of worry from you — where there is reticence, transform it to caution and seek then wisdom."

I realized at that point, angels have little understanding of, or perhaps they do not even see, all the petty foibles and emotional pressures with which we humans have to deal. Possibly these doubts and fears do not even register on their 'thought-screens.' I wondered whether we had to deal with these levels of mental activity ourselves or whether, in future situations, we could simply ask the angels to lift them off us. Talantia addressed the question without being asked.

"Where there are concerns for the self that are from our point of view illusory concerns, we would seek to move in the mind of the seeker through such channels as have been opened in the act of seeking. To cleanse these channels in ways both delightful and delicate, gentle and harmonious. To affirm within that there is joy in pursuit of wisdom. And further, our presence is to take these illusions which would be of the material dimensions and, as dust upon the glorious garment, which is the human form, with gentleness of our touch, show that this dust may be wiped away.

"There is no shaking in this. There is no ritualistic ablution. There is no abasement. There is only the gentle, loving touch of friend to friend to say "Here is a thing you need not carry." It may be wiped away as a mote. Things of no significance to the spiritual reality, which is the emergent reality, may be released in this way."

The atmosphere in the room seemed to kick into a new gear. The dusting of the garments, which I imagined some large but gentle invisible hands wafting clean, left the air glisteringly clear. We were being drawn into their closeness and caring.

Talantia continued, honing in on the crux of the fear.

"This shift of consciousness would be away from the fiercesome; there are sufficient ideations of an illusory deity of fiercesome nature. The world-mind abounds in these. Given these are creations of man, he would see them as having value still. When he sees they have diminishing value, in that the return sought from such images is not the true, nor in essence, a beautiful vision, then there is a falling away of these ideations.

"In the contentions for form, the thundering and the shaping are not of the Spirit. These are of the minds of man. The choice in the mind of man is to release these or carry them forward. Here we cannot choose. Here we must only abide or assist, given the nature of the choices made by mortals.

"In the essence of the message of everlasting love and renewed hope, humans may find a great desire for the new. Then you will see a flurry of activity within the human realms as they would seek, in their own ways, to make clean the vessels and to beautify their environments — and all this would be an illusion of making themselves worthy to receive.

"We emphasize that it is illusion, for you are fully worthy to receive as you are now. Yet, there is that energy which says "I must prepare — I must be ready." The value in such a flurry of activity is to prepare the mind for an acceptance of its own readiness. Happily, when the material realm appears ready, so is the mind.

"Imbibe at the fountains of peace profound; luxuriate in the wealth bestowed on all; attune to the flow of abundance. Here is the joy of existence — know there is enough. There is a plenitude and it exists for all. In your finite terms, the infinity of God's love is inexhaustible. This is the reality."

At this juncture Edward motioned for a break.

The feeling in the room meanwhile had become one of gentle conversation. We were all relaxing into friendship. For me the accent was on the penetrating wisdom revealed in the encounter. We were starting to understand the nature of the intelligence of these entities. The more reality and credence we gave the communication and the angels, the more pronounced the spirit feelings became. The doubts and worries dissolved — the first part of the session had taken care of them — and we were all getting along famously. At that stage we had little idea of what lay ahead.

We were talking about the names the angels took, wondering why, and whether they had a particular meaning for them. Edward had told us the angels used signatures, as they had called them, glyphs that resolved into spelt names on his 'thought-screen'; this was the way the angels introduced themselves. When we resumed transmission, Durandior elaborated.

"A name is a definition of a discrete energy form and an aspect of individuated mind that allows direct communication from the one to the other . . .

"Now for our purposes, we would choose names in the phonemes of the language of the realm which might in ways reflect some aspects of ourselves to you, that you may know us in a fuller sense.

"We are minded beings. We have personhood not exactly as you would understand it, and yet it is so. We have aspects of our mindedness and of our being which we cannot, alas, at this time and juncture, fully share with you. But Oh! there are glories that await us all as you ascend your joyful growth. Of this we are certain and of this you may be sure.

"Now, as to the reality content of the names assigned, the signatures. These are only valuable insofar as they may communicate information to you. You may reach us however, through other vibrations where name, or individuated mind, or discrete personhood, are not necessary for the communion to take place. Here is a phenomenon of externalization, for projection into an outer Universe of an inter-dimensional reality."

We asked her the meaning of Durandior.

"Within my essence there is a golden light," she replied fluidly, "hence the ending 'dior.' That in my energy there is an exhilaration in pursuit and discovery of knowledge. This is an area of my great love. Here I am 'running' to my 'destiny.' I would then be Durandior." The angel finished and withdrew amidst our laughter at the way she had put it all together.

We wondered about Talantia and, within moments, another softer and more serious voice came through Edward.

"I, Talantia, step forward and may transcribe this for you. To foster progress is to bring forth talents. Then I would be one who would do this."

She went on to draw our attention to the parable of the faithful stewards, emphasizing as she had previously, the immense potential in the consciousness of stewardship.

Next, it was Beatea. No prizes for guessing that one immediately. "Beatitude?"

She sounded delighted. "Indeed, you see now the enlightenment is of beatific vision. To foster this I would embody the beautiful. Hence, I am Beatea."

After her, Karyariel. She likened herself to a vehicle for moving forward into the future. "In our joyous energy and exuberance, I may be chosen by the name Karyariel."

Her tone changed and became more businesslike. She informed us there were representatives of all twelve angelic groups, as we would understand it, present. We realized with astonishment, the entire planetary host of angels, who are organized into twelve divisions of specific ministry, were interested in this unusual gathering, members of each expressing their desire to come forward. Karyariel indicated the choice was up to us. A quite impossible task — we loved them all and would not know where to start choosing.

Durandior came to our rescue and introduced Petrous, an Angel of the Churches. We knew this group had long been on the planet and were the main sponsors of much of its traditional spiritual activity.

"Look back upon the preceding two millennia," Petrous urged us, "and see with amazement, as we would see, the nature of the changes that have been wrought upon this world."

She was not blind to the tangents and interminable digressions shown by the churches, placing the responsibility for the confusion on a striving she perceived in us humans, to imbue a single institution with all authority. She added quickly that there was a need for this in the overall design, and the bringing forward of the order was useful and necessary.

This was hitting near the bone for me since I have frequently railed at the excesses of the traditional churches. What she was suggesting was a far more loving and gentle approach.

"Mortals, human brethren," she had said, "as you are alive today and would look back upon your history, look back in blessing to those who build the way for you. Condemn them not for their errors; rejoice in the Truth they brought forward.

"See what a wonderment would be unveiled for you leaving behind recriminations. Leave behind the world of old forms and by looking to past, present and future in love and blessing, you will find our way. The Heavenly Way.

"The good is always. The good is where you may focus your mindedness. Truth, Beauty and Goodness. These are not mysteries to your experience, they are verities. Here you may truly **know**."

We were all moved by the ease and integrity of what the angel told us. What good could come out of ranting on about the horrors of the Inquisition or the hypocrisy of religion as big business? It would not change the past one iota and organized religion is quite evidently losing its hold on those of our generation.

Then came a revealing, if inadvertent, insight into the nature of the angelic mind. Petrous had just drawn to a close an extended monologue on churches and the cathedral building era. These acts of devotion, she saw as marking the great movements of Spirit over the land. She finished by expressing her angelic confusion at what was currently transpiring in these buildings. With a sweetness and innocence she added, "We wonder at this. We say, 'How is it that mortals would choose this?' In this, our Mother says, 'Be at peace, my children, for here are your younger brethren learning,' and She takes us back to the days of our youth, when, in joyful exuberance we saw much but understood little."

I remembered how, in that earlier session, the angel had quietened our impatience with much the same wisdom. I saw what a wholly unpunitive viewpoint they hold. They are always encouraging; always seeing the best in people and events. Not, I think, entirely blind to our shortcomings — they had shown that by lifting the worries off us — but inevitably moving us towards seeing the **Divine** Causes. Seeing the underlying plan.

Elyan was the next to introduce herself. One of the Home Seraphim and a representative of a working unit evidently among the first to function on this world. She reminded us that a home is not a house.

"The home is the ambience," she had said, "the energy forms and collective love which act as a reservoir upon which great drawing power is rested by the Divine."

She noted the transitional nature of the times reflected in the diminishment of the importance of the home as a place offering succor and sanctuary. This was the function of the home, she said. "The center, the peaceful place. One may move out from it, but there is always the promise of the return. See here the wisdom of the Father in the portrayal of eternal reality; for you have moved out from your home and will return to it before the end of time."

She talked about home-makers being the stabilizers of the world, and within the context of the spiritual center of our lives, moving from the churches, back into our hearts and hearths. I was starting to appreciate in a new way, the value of the home in the great changes to come.

Elyan elaborated: "See in your minds the home as a point of light shining in a darkened world. The light moves out and interconnects with the light of other homes. And in this way a network of light stands durable . . . and the simplicity of spirituality may be revealed."

The angels had seen the Father's Will, she maintained, and that was that all families should be housed graciously. She praised the great strength of the mortal woman who kept the importance of the home alive throughout the commotions and perturbations of our planet's turbulent history.

She ended her presentation with an evocative intimation of another form of life as yet only dimly revealed to us — the elves and the nature spirits.

"I thank you for the opening . . . and will be glad to

come again. Before I leave I would add that we work as well with other angelic intelligences not denominated among the twelve, but revealed from other sources. Here we speak of our cousins who foster the gardens. For in truth, the home and the garden are all God's vision and His will for His children."

With that it was the turn of an Angel of Health. It was a simple message. The key to health lay in happiness. She acknowledged the new movement towards wholism, commenting about it that "it sees today with the wisdom of an ancient; that a human body also has a spiritual being needing to be ministered to."

Her definition of spiritual health I found particularly pertinent, as I sometimes indulge in the very error she focussed upon. The 'grass is greener' syndrome seems to be a piece of misinformation which goes with the human equipment.

"Spiritual health," she had said, "is doing with joy, the work of the dimension in which a being finds itself," and the Angel of Health then gave us a visual simile with which to illustrate it. An image of a Being on a beach, waiting to transform itself into a winged creature to fly away to the promised land where things are seen as 'better.' "Things cannot be better far away," she had completed the thought. "This is an error. A Being is whole and holy the moment it sees itself at peace in God, its loving Parent. Happiness, health, joy and resolute action ensue from this."

Durandior, ever vigilant as to our well-being, must have noticed the first signs of strain, though she expressed herself with her customary grace. "The collective energies here are approaching a level of surfeit," she said, suggesting we chose only two more from the angels we were told had foregathered.

We asked if there were any Angels of Education

accessible? A short silence later. "Call me Mentoria," and we got another glimpse into angelic affairs.

"Our new mandate," she was saying "is to enjoin with the Home Seraphim and the Angels of Progress." They had in mind an education which would once again be carried out by parents rather than a 'teaching elite.'

"And in the upbringing of their children," she reported, "the parents would decide what is of value and what is to be fostered. Here you may see how rapidly the new culture may emerge in your midst."

She pointed to new technology, soon to be more generally available, that would decentralize knowledge. She saw a time when schoolteachers, as we know them, would be "without suitable institutional employment." When they may "come to work in cooperation with others in a symbiosis which transcends the commerce of money in the energy exchange." True teachers would be among us again when there started being "an exchange of information for shelter, sanctuary and community."

She emphasized the importance, not surprisingly, of the spiritual education of the young of today. The angels had received notice from their "higher levels" that "great ones will come through these souls." She ended her transmission with a most impassioned appeal.

"What is of greatest value in the education of these young ones is that they look to nature. To find therein the patterns, placed by Intelligences Divine, and reflecting the eternal realities and the unfolding nature of time and space.

"Then we may come through to offer advancements, elaborations and insights which will accelerate the process and bring spiritual realities unknown on this world.

"What is in this is the decline of all forms of violence, and an ascendance of all forms of loving understanding, of compassionate outreach, of sharing, spiritual, material and mindal. You would then see among these new ones a fuller portrayal of the potentials long-held within the mortals of the realm.

"Joy is the essence of our mission," she finished, "joy and the satisfaction of the unfolding of the plan."

One more angel to go. Who better, we thought, than one of the Angels of Entertainment! And, indeed, it was an inspiring message which I have reproduced in full. It has a key for all of us as we start wondering what we might actually **do** to further God's plan. The angel did not give her name; here is what she told us.

"And now, see before you the joys of existence you will all find in the artistic creations which are your work. You have chosen this. It is, in truth, through these channels that the redemption may be most fully made known.

"Other corps and legions (of angels) have perforce been obliged to work in areas tainted by confusion. Here I would see the Angels of Nation Life, those concerned with political machinations; the Angels of Education and of Health, all wearing, as it were, leaden boots. Yet among us, a greater freedom; among us, a creative expression.

"Indeed, at times, those who are illusorily empowered have made difficult this free expression, for they well understood a threat to their own power base. With a release from fear, with joyous action, with exhilaration . . ." the angel was clearly groping, "there isn't a word in this channel to express what we would seek . . .

"In essence, it is the freedom to express the Divine within which has been greatly liberated, highly upstepped. Here you might understand the Master would seek to move through those who would invite Him to share their minds.

"Then would our beloved Master find a fuller expression for that which was denied Him in His material bestowal. Here you may find His energies of delight and love more fully manifest in yourselves. The power with which this will speak to all the world should not be understated.

"This is no less than the long-hoped for spiritual rebirth among the arts. The new renaissance as has been forecast in the minds of many. This is what you have sought; this is what you have found.

"Now you would see within our new mandate, we are enjoined to foster laughter . . ."

And an astonishing thing happened. It was like being tickled on the most subtle levels of our beings. Laughter was irresistible; we started roaring our heads off. There was no holding it back. Tears rolled down our faces.

"You are asked to see how the joy of laughter will set the world a-turning."

We were all caught up in the wonderful joke. Laughter coming from the furthest and deepest places in the Universe. And we had taken it all so seriously . . .

Contact that had started uptight and worried had come together in this magnificent, shared humor. We rolled around joyously, barely able to hear the angel's last words.

It was in this vastly opened state, when all of us thought the session was over, the angels retired for the evening, that the last miraculous contact burst through.

"I am Shandron." A new voice rolled out of Edward, deep and resonant, falling in waves around us.

"I am a Being of advanced status, greater than Seraphim. I am one who would be of Superuniverse status. I am Supernaphim."

A sense of quiet awe descended on the room. The

presence of this entity was incontrovertible. It was as though we were receiving a Universe bulletin. An update from the furthest reaches.

The voice continued inexorably.

"My place here in this Great Work is as an usher of this new dispensation that is upon us, and I emphasize this aspect.

"Yes, there has been a turning of a dispensation for your world. An adjudication, in a sense. A release from patterns long-held. But more, and this you know in your hearts."

If I had any remaining doubts about the decisions pertaining to the Lucifer Rebellion, they were rapidly disappearing as this majestic voice moved on.

"Such are the nature and magnitude of the changes being wrought here on this world that we, of Superuniverse status, are invited to function with our younger sisters and brethren in the ordering of the new ways.

"This I have to say on the nature of epochal transformations. You would do well to see that the gradation of change has been accelerated. There are moments in the previous century when dawn lights shone in minds. These minds, brought forth from the land called England, found correspondences in the land called America. There was the philosophy of the Divine Transcendence and in the movement of man to a fuller realization of his status.

"A channel was opened for the mentation of the realm. Here I speak in America of those known as the transcendental philosophers. Look to an understanding of the world views of Emerson, Thoreau and Whitman. Here were seeds planted that made way for the greater revelations that are at hand.

"In this upliftment were there born many spiritual

movements of mind. Many of these were ritualistic, couched in the old forms. They served the purpose of moving the energies which were offered, into the minds of men. And yet we are glad to see these older forms have fallen away to a great extent.

"As to the nature of the epochal revelation you contain. And here I do not refer to the book on the history and relationship of this world to the Universe of Universes. No indeed, not this, for this is but a step to the fuller realization. The new dispensation upon us all, in the light of the Supreme and His great movements, flowing out from the Center Universe, going into many forms and flowing through many minds, this new dispensation is coming to reveal and be revealed here and in eternity. In its fullness, it will require the completion of time to be understood by all. What is given to you in your outworking, in your forward motion in the ascension through higher worlds, are the crystalline insights into the eternal verities, given to those youngest and yet with the hearts to feel and the minds to understand.

"You would do well to understand that the old things have, indeed, passed away; that **all** things are made new. It is not the end of revelation to this world, it is **your** lives that are the Supreme revelation. It is the Master Son moving through you that is the wonder of the Universe.

"These things cannot be contained by any single entity nor by any established group. These things are for all beings to join in worshipful wonder of the Father, who, in His wisdom and mercy, would create creatures so lowly, and yet ones that may be raised up into His very presence in an instant of time.

"Praise then, this God of whom we are all children. This is the wonder of our Supreme Being.

"We work to bring in new levels of organization. This will be a gentle transfer and upliftment. Much that is held in the mind as it exists today are but sketches on a note pad. These sketches will allow for the furtherance of the vision of the totality of the change. But it is in the living experience that the change may be made real.

"Leave then your studies. Leave then your books. Leave all these things into their proper place in your lives. Find not your center in them, but within yourselves. Live your lives as they have been given you and in **this** will be the wonder for all to see. We carry forward the message, we carry forward the vision, we carry forward the glorious spectacle unfolding on this wonderworld, this place of peace. It is this planet of which I speak that is held in most loving concern by **all** the Universe of Universes, that in God's mercy **all** may live in bliss and harmony.

"Now we speak in this epic, of the redeemed ones. Great then is the wisdom and mercy of the Father, and such is His love that none are lost. All have been found and will stand, revealed with gowns of innocence. For full forgiveness has been lodged within the hearts of simple creatures and, moved by this vision, the High, the Mighty, as you would see them, see a revelation of their most revered Father.

"That such lowly creatures may love so greatly is, and ever shall be, the message for all to learn. Forgive the error, see the Son, enjoin as brothers, be us all as one.

"This is the ending of my message for you. I will greet you again."

AFTERWORD

The events recorded in this book all occurred in the first two years of the 1980's, but the sequence which so evidently developed has continued unabated to this day.

I pondered long as to whether to bring them to the notice of my fellow planetary citizens and in what form they might be most readily understood. As anyone might appreciate, the matters are so intensely personal and thus fly in the face of current standards of objective verification. However the two intervening years have for me continued to sustain the veracity of the basic issue. That there has indeed been a reconciliation between conflicting elements within the larger matrix of Universe affairs now seems to be borne out by the increasing pace with which our own transmutations and transformations are happening.

It is becoming more evident to us that our small planet is considered the pearl of this area of the Universe and its demise amid atomic devastation is simply not being allowed to occur. This is not an act of coercion or a diminishment of our free will by those whose function it is to watch over Universe affairs, since no human being truly desires in his heart to see their planet destroyed. It is the long-awaited helping hand for which we have all so ardently hoped and prayed. The presence of these entities from other levels and domains of the Master Universe does not usurp our prerogatives as Will creatures but seeks to help us mend a situation which has been regarded as immensely difficult for the last two hundred thousand years, and which happened largely through no fault of our own. The extra-terrestrials and the angels are among those here to aid any who would extend a hand to them and, though their presence will become more evident as the years pass, they will continue to devote their efforts to the more subtle levels of transformation.

The Angels have made it clear that we should not anticipate overt help in the form of any single figure, but rather seek to encourage a collective group-messiah along the lines of a Network of Light of World Servers who, by their efforts and hearts' desires, will transmute the energies coursing through our world at this point in time.

This entity, a group soul or Anima Mundi, appears to be composed of all sentient intelligence systems on the planet working together in some form of transcendent unity. We are all a part of this to the degree each of us can sustain the vision, and the lessons of our interdependence lie in allowing ourselves to support and affirm each other's unique contribution to this growing Horus Child of the planet — a child who truly grows in the heart of each one of us, mortal, extra-terrestrial and angel alike.

It is this linkage which each and every one of us will start to experience as we are brought, or bring ourselves, into synchrony with the Master Plan. We discover through our own experimentations those factors which benefit **all** participants, and gradually bring our lives into "meaningful coincidence" with the best of our intentions. This is the Kingdom of Heaven, the New Jerusalem and the New Aeon brought into manifestation from the Spirit reality into the world of matter. It is a concept which lies deep in all the major planetary belief systems and which, by its actual eventuation, will bring about healing of the fragmented mentation of the realm.

It is however for each of you reading this to allow the reality of the Reconciliation of the Rebellion to percolate in your souls. We all have the innate capacity to discern Truth when we **feel** it and, as the little boy said, it is **feelings** that all intelligent species have in common.

I am closing this phase of the adventures with a transmission received in Canada by Edward on the Vernal Equinox of 1982. The words came through in a flood of automatic writing and for me they bear the undeniable stamp of authority of the Being who lives in my heart. Although we did not come to piece the jigsaw together until well after the events, Edward received this transmission on the very night my companion and I stood on that little bridge in the Bahamas and knew, with such intuitive intensity, **"The Circuits are Opening! The Circuits are Opening!"**

A MESSAGE FROM MICHAEL

Here is the promise kept and the vision offered for your greater understanding. For whosoever will gather together with purity of love and sincerity of desire to know me, there shall I be with them to offer my love, my comfort, my peace.

"You know me for I am within you. I am your creator-father and your brother. I am Michael. Peace.

"Standing in your midst in spirit, I come to assure you that through the fiery trial faced by your world, I will protect those who know me. Those who have decided against truth, peace, and all that ensues from the Godly nature shall

pass before your eyes; this is Divine Justice and the Law.

"Choice is the essence of my message. You are free to choose your own way, free to choose life and free to deny life. I would have you choose to know me and, through me, our great Father, but like Him, I will never force your choice. Understand this and you will approach a fuller understanding of the reasons for the suffering and the horror that you meet on this troubled world.

"To ask me what you must do at this time or at any other, is to ask me to choose for you. Although I may choose you and love you and protect you as my precious children, I also respect your selves and so would not do this. Yet the true question you pose is one that is foremost in the minds of many at this time; "What can I do to help God's plan and bring peace?"

"It is most natural and most beautiful that you should seek to preserve the integrity of your home planet; I join with you in your love for the wondrous beauty that has evolved from our plans for Urantia. I too **will** peace for the world of my mortal bestowal and should this not tell you that peace will be?

"When you look at your world, what do you see? Given you would see what you would choose to see, what have you chosen?

"The call sounded by angels and Sons of God is to see the vision of Light and Life; your work may be then, to dare to envision planetary peace. Within you is such a power that only God, our Father, may safely contain it. He holds that power for you until you choose to wield it, and it may only be used to effect His Will and obey His Laws. If you would come to understand that His Will is Life and His Law is Love, you would find the keys to unlock the Universe of power and majesty within yourselves.

"You have the ability to rest this night and awake tomorrow as a realizing Son of God. You have the ability to so live your life according to the faith I have shared with you that men and women would know of a certainty that the Sons and Daughters of God are abroad on the face of the earth. To do so is to stand steadfastly for the truth in such a gentle way, that it may shine forth from your eyes and awaken those who walk in their dream of fear.

"What can you do to be as such? Simply love one another as I love you. All else will flow from the rivers of Divine Creativity that would be released if you but dared to do this simple, natural and inherent act.

"Remember, my beloved friends, that Love is the Power of the Universe. This is the lesson I would have you learn perfectly."

Coleman Publishing
99 Milbar Boulevard
Farmingdale, New York 11735
(516) 293-0383-84